Organic Farming

Organic Farming

Donnie Stallone

SYRAWOOD
PUBLISHING HOUSE

New York

Published by Syrawood Publishing House,
750 Third Avenue, 9th Floor,
New York, NY 10017, USA
www.syrawoodpublishinghouse.com

Organic Farming
Donnie Stallone

© 2022 Syrawood Publishing House

International Standard Book Number: 978-1-64740-064-4 (Hardback)

Cataloging-in-Publication Data

Organic farming / Donnie Stallone.
    p. cm.
Includes bibliographical references and index.
ISBN 978-1-64740-064-4
1. Organic farming. 2. Agriculture. I. Stallone, Donnie.
S605.5 .O74 2022
631.584--dc23

Table of Contents

Preface

This book is a culmination of my many years of practice in this field. I attribute the success of this book to my support group. I would like to thank my parents who have showered me with unconditional love and support and my peers and professors for their constant guidance.

Organic farming is an alternative agricultural system that focuses on the use of fertilizers that are organic in nature such as compost manure, bone meal and green manure. It advocates the use of farming techniques such as crop rotation and companion planting. Organic farming also promotes biological pest control, fostering of insect predators and mixed cropping. It particularly focuses on the use of naturally occurring substances and strictly eliminating the use of synthetic substances. The farming methods within organic farming use a combination of ecology and modern technology along with traditional farming practices which are based on naturally occurring biological processes. This book studies, analyses and upholds the pillars of organic farming and its utmost significance in modern times. It brings forth some of the most innovative concepts and elucidates the unexplored aspects of this field. Those in search of information to further their knowledge will be greatly assisted by this book.

The details of chapters are provided below for a progressive learning:

Chapter – What is Organic Farming?

Organic farming is an alternative agricultural system that encourages the use of the fertilizers of organic origins such as green manure, compose manure and bone meal. It focuses on sustaining and enhancing the health of soil. This is an introductory chapter which will introduce briefly all the significant aspects of organic farming.

Chapter – Methods and Practices in Organic Farming

The most common methods and practices of organic farming are crop-rotation, vegan-organic gardening and composting. These measures make use of the natural environment to enhance productivity. The topics elaborated in this chapter will help in gaining a better understanding of these practices and methods of organic farming.

Chapter – Diverse Types of Organic Fertilizers

The fertilizers derived from animal matter, vegetable matter, animal excreta and human excreta are known as organic fertilizers. They can be used as a substitute of chemical fertilizers. This chapter has been carefully written to provide an easy understanding of the varied facets of organic fertilizers as well their advantages and disadvantages.

Chapter – Soil Management in Organic Farming

Soil management includes the application of treatments, operations and practices to protect soil and enhance its performance. Soil management in organic farming focuses on using soil health as the major source for fertilization and pest control. All the techniques and methods related to soil management in organic farming are introduced in this chapter.

Chapter – Organic Pest and Disease Management

Organic pest and disease control includes diverse methods to manage pests and diseases such as biological pest control, organic weed control, cultural weed control and early blight management. This chapter discusses in detail these methods and practices related to organic pest and disease management.

Donnie Stallone

1

What is Organic Farming?

Organic farming is an alternative agricultural system that encourages the use of the fertilizers of organic origins such as green manure, compose manure and bone meal. It focuses on sustaining and enhancing the health of soil. This is an introductory chapter which will introduce briefly all the significant aspects of organic farming.

Organic farming is the agricultural system that uses ecologically based pest controls and biological fertilizers derived largely from animal and plant wastes and nitrogen-fixing cover crops. Modern organic farming was developed as a response to the environmental harm caused by the use of chemical pesticides and synthetic fertilizers in conventional agriculture, and it has numerous ecological benefits.

Compared with conventional agriculture, organic farming uses fewer pesticides, reduces soil erosion, decreases nitrate leaching into groundwater and surface water, and recycles animal wastes back into the farm. These benefits are counterbalanced by higher food costs for consumers and generally lower yields. Indeed, yields of organic crops have been found to be about 25 percent lower overall than conventionally grown crops, although this can vary considerably depending upon the type of crop. The challenge for future organic agriculture will be to maintain its environmental benefits, increase yields, and reduce prices while meeting the challenges of climate change and an increasing world population.

Regulation

Organic agriculture is defined formally by governments. Farmers must be certified for their produce and products to be labeled "organic," and there are specific organic standards for crops, animals, and wild-crafted products and for the processing of agricultural products. Organic standards in the European Union (EU) and the United States, for example, prohibit the use of synthetic pesticides, fertilizers, ionizing radiation, sewage sludge, and genetically engineered plants or products. In the EU, organic certification and inspection is carried out by approved organic control bodies according to EU standards. Organic farming has been defined by the National Organic Standards of the

U.S. Department of Agriculture (USDA) since 2000, and there are many accredited organic certifiers across the country.

Although most countries have their own programs for organic certification, certifiers in the EU or the United States can inspect and certify growers and processors for other countries. This is especially useful when products grown organically in Mexico, for example, are exported to the United States.

Organic Farming Methods

Fertilizers

Since synthetic fertilizers are not used, building and maintaining a rich, living soil through the addition of organic matter is a priority for organic farmers. Organic matter can be applied through the application of manure, compost, and animal by-products, such as feather meal or blood meal. Due to the potential for harbouring human pathogens, the USDA National Organic Standards mandate that raw manure must be applied no later than 90 or 120 days before harvest, depending on whether the harvested part of the crop is in contact with the ground. Composted manure that has been turned 5 times in 15 days and reached temperatures between 55–77.2 °C (131–171 °F) has no restrictions on application times. Compost adds organic matter, providing a wide range of nutrients for plants, and adds beneficial microbes to the soil. Given that these nutrients are mostly in an unmineralized form that cannot be taken up by plants, soil microbes are needed to break down organic matter and transform nutrients into a bioavailable "mineralized" state. In comparison, synthetic fertilizers are already in mineralized form and can be taken up by plants directly.

Soil is maintained by planting and then tilling in cover crops, which help protect the soil from erosion off-season and provide additional organic matter. The tilling in of nitrogen-fixing cover crops, such as clover or alfalfa, also adds nitrogen to the soil. Cover crops are commonly planted before or after the cash crop season or in conjunction with crop rotation and can also be planted between the rows of some crops, such as tree fruits. Researchers and growers are working to develop organic farming "no-till" and reduced-tillage practices in order to further reduce erosion.

Farmer managing a compost pile.

Pest Control

Organic pesticides are derived from naturally occurring sources. These include living organisms such as the bacteria Bacillus thuringiensis, which is used to control caterpillar pests, or plant derivatives such as pyrethrins (from the dried flower heads of Chrysanthemum cinerariifolium) or neem oil (from the seeds of Azadirachta indica). Mineral-based inorganic pesticides such as sulfur and copper are also allowed.

In addition to pesticides, organic pest control integrates biological, cultural, and genetic controls to minimize pest damage. Biological control utilizes the natural enemies of pests, such as predatory insects (e.g., ladybugs) or parasitoids (e.g., certain wasps) to attack insect pests. Pest cycles can be disrupted with cultural controls, of which crop rotation is the most widely used. Finally, traditional plant breeding has produced numerous crop varieties that are resistant to specific pests. The use of such varieties and the planting of genetically diverse crops provide genetic control against pests and many plant diseases.

Differences between Organic and Conventional Farming Methods

In conventional farming method, before seeds are sown, the farmer will have to treat or fumigate his farm using harsh chemicals to exterminate any naturally existing fungicides. He will fertilize the soil using petroleum based fertilizers. On the flip side, the organic farmer will prepare and enrich his land before sowing by sprinkling natural based fertilizers such as manure, bone meal or shellfish fertilizer.

Before planting seeds, the organic farmer will soak the seeds in fungicides and pesticides to keep insects and pests at bay. Chemical are also incorporated in the irrigation water to prevent insects from stealing the planted seeds. On the other hand, the organic farmer will not soak his seeds in any chemical solution nor irrigate the newly planted seeds using water with added chemicals. In fact, he will not even irrigate with council water, which is normally chlorinated to kill any bacteria. He will depend on natural rain or harvest and stored rainwater to use during dry months.

When the seeds have sprung up, and it's time to get rid of weeds, the conventional farmer will use weedicide to exterminate weeds. The organic farmer will not use such chemicals to get rid of the weed problem. Instead, he will physically weed out the farm, although it's very labor intensive. Better still, the organic farmer can use a flame weeder to exterminate weeds or use animals to eat away the weeds.

When it comes to consumption, it's a no-brainer that anyone consuming products from the conventional farmer will absorb the pesticide and weedicide residues into the body, which could lead to developing dangerous diseases like cancer. People understand that health is important to them and that's why they are going organic in record numbers today.

Reasons for Organic Farming

The population of the planet is skyrocketing and providing food for the world is becoming extremely difficult. The need of the hour is sustainable cultivation and production of food for all. The Green Revolution and its chemical based technology are losing its appeal as dividends are falling and returns are unsustainable. Pollution and climate change are other negative externalities caused by use of fossil fuel based chemicals.

In spite of our diet choices, organic food is the best choice you'll ever make, and this means embracing organic farming methods. Here are the reasons why we need to take up organic farming methods:

To Accrue the Benefits of Nutrients

Foods from organic farms are loaded with nutrients such as vitamins, enzymes, minerals and other micro-nutrients compared to those from conventional farms. This is because organic farms are managed and nourished using sustainable practices. In fact, some past researchers collected and tested vegetables, fruits, and grains from both organic farms and conventional farms.

The conclusion was that food items from organic farms had way more nutrients than those sourced from commercial or conventional farms. The study went further to substantiate that five servings of these fruits and vegetables from organic farms offered sufficient allowance of vitamin C. However, the same quantity of fruits and vegetable did not offer the same sufficient allowance.

Stay Away from GMOs

Statistics show that genetically modified foods (GMOs) are contaminating natural foods sources at real scary pace, manifesting grave effects beyond our comprehension. What makes them a great threat is they are not even labeled. So, sticking to organic foods sourced from veritable sources is the only way to mitigate these grave effects of GMOs.

Natural and Better Taste

Those that have tasted organically farmed foods would attest to the fact that they have a natural and better taste. The natural and superior taste stems from the well balanced and nourished soil. Organic farmers always prioritize quality over quantity.

Direct Support to Farming

Purchasing foods items from organic farmers is a surefire investment in a cost-effective future. Conventional farming methods have enjoyed great subsidies and tax cuts from most governments over the past years. This has led to the proliferation of commercially produced foods that have increased dangerous diseases like cancer. It's time governments invested in organic farming technologies to mitigates these problems and secure the future. It all starts with you buying food items from known organic sources.

To Conserve Agricultural Diversity

These days, it normal to hear news about extinct species and this should be a major concern. In the last century alone, it is approximated that 75 percent of agricultural diversity of crops has been wiped out. Slanting towards one form of farming is a recipe for disaster in the future. A classic example is a potato. There were different varieties available in the marketplace. Today, only one species of potato dominate.

This is a dangerous situation because if pests knock out the remaining potato specie available today, we will not have potatoes anymore. This is why we need organic farming methods that produce disease and pest resistant crops to guarantee a sustainable future.

To Prevent Antibiotics, Drugs, and Hormones in Animal Products

Commercial dairy and meat are highly susceptible to contamination by dangerous substances. A statistic in an American journal revealed that over 90% of chemicals the population consumes emanate from meat tissue and dairy products. According to a report by Environmental Protection Agency (EPA), a vast majority of pesticides are consumed by the population stem from poultry, meat, eggs, fish and dairy product since animals and birds that produce these products sit on top of the food chain.

This means they are fed foods loaded with chemicals and toxins. Drugs, antibiotics, and growth hormones are also injected into these animals and so, are directly transferred to meat and dairy products. Hormone supplementation fed to farmed fish, beef and dairy products contribute mightily to ingestion of chemicals. These chemicals only come with a lot of complications like genetic problems, cancer risks, growth of tumor and other complications at the outset of puberty.

Key Features of Organic Farming

1. Protecting soil quality using organic material and encouraging biological activity.

2. Indirect provision of crop nutrients using soil microorganisms.

3. Nitrogen fixation in soils using legumes.

4. Weed and pest control based on methods like crop rotation, biological diversity, natural predators, organic manures and suitable chemical, thermal and biological intervention.

5. Rearing of livestock, taking care of housing, nutrition, health, rearing and breeding.

6. Care for the larger environment and conservation of natural habitats and wildlife.

Four Principles of Organic Farming

1. Principle of Health: Organic agriculture must contribute to the health and well being of soil, plants, animals, humans and the earth. It is the sustenance of mental, physical, ecological and social well being. For instance, it provides pollution and chemical free, nutritious food items for humans.

2. Principle of Fairness: Fairness is evident in maintaining equity and justice of the shared planet both among humans and other living beings. Organic farming provides good quality of life and helps in reducing poverty. Natural resources must be judiciously used and preserved for future generations.

3. Principle of Ecological Balance: Organic farming must be modeled on living ecological systems. Organic farming methods must fit the ecological balances and cycles in nature.

4. Principle of Care: Organic agriculture should be practiced in a careful and responsible manner to benefit the present and future generations and the environment.

As opposed to modern and conventional agricultural methods, organic farming does not depend on synthetic chemicals. It utilizes natural, biological methods to build up soil fertility such as microbial activity boosting plant nutrition.

Secondly, multiple cropping practiced in organic farming boosts biodiversity which enhances productivity and resilience and contributes to a healthy farming system. Conventional farming systems use mono cropping that destroys the soil fertility.

Organic farming at backyard.

Why is Modern Farming Unsustainable?

1. Loss of soil fertility due to excessive use of chemical fertilizers and lack of crop rotation.

2. Nitrate run off during rains contaminates water resources.

3. Soil erosion due to deep ploughing and heavy rains.

4. More requirement of fuel for cultivation.

5. Use of poisonous bio-cide sprays to curb pest and weeds.

6. Cruelty to animals in their housing, feeding, breeding and slaughtering.

7. Loss of biodiversity due to mono culture.

8. Native animals and plants lose space to exotic species and hybrids.

Benefits of Organically Grown Food Items and Agricultural Produce

1. Better Nutrition: As compared to a longer time conventionally grown food, organic food is much richer in nutrients. Nutritional value of a food item is determined by its mineral and vitamin content. Organic farming enhances the nutrients of the soil which is passed on to the plants and animals.

2. Helps us stay healthy: Organic foods do not contain any chemical. This is because organic farmers don't use chemicals at any stage of the food-growing process like their commercial counterparts. Organic farmers use natural farming techniques that don't

harm humans and environment. These foods keep dangerous diseases like cancer and diabetes at bay.

3. Free of poison: Organic farming does not make use of poisonous chemicals, pesticides and weedicides. Studies reveal that a large section of the population fed on toxic substances used in conventional agriculture have fallen prey to diseases like cancer. As organic farming avoids these toxins, it reduces the sickness and diseases due to them.

4. Organic foods are highly authenticated: For any produce to qualify as organic food, it must undergo quality checks and the creation process rigorously investigated. The same rule applies to international markets. This is a great victory for consumers because they are getting the real organic foods. These quality checks and investigations weed out quacks who want to benefit from the organic food label by delivering commercially produced foods instead.

5. Lower prices: There is a big misconception that organic foods are relatively expensive. The truth is they are actually cheaper because they don't require application of expensive pesticides, insecticides, and weedicides. In fact, you can get organic foods direct from the source at really reasonable prices.

6. Enhanced Taste: The quality of food is also determined by its taste. Organic food often tastes better than other food. The sugar content in organically grown fruits and vegetables provides them with extra taste. The quality of fruits and vegetables can be measured using Brix analysis.

7. Organic farming methods are eco-friendly: In commercial farms, the chemicals applied infiltrate into the soil and severely contaminate it and nearby water sources. Plant life, animals, and humans are all impacted by this phenomenon. Organic farming does not utilize these harsh chemicals so; the environment remains protected.

8. Longer shelf–life: Organic plants have greater metabolic and structural integrity in their cellular structure than conventional crops. This enables storage of organic food for a longer time.

Organic farming is preferred as it battles pests and weeds in a non-toxic manner, involves less input costs for cultivation and preserves the ecological balance while promoting biological diversity and protection of the environment.

Importance of Organic Farming

Modern farming practices heavily based on use of chemical fertilizers and pesticides, have created problems of land degradation, environmental pollution, deforestation, biodiversity depletion, seepage and water logging, lowering of ground water tables, inter crop disparities, emergence of several diseases, pest multiplication, pest resurgence and resistance.

Modern agricultural practices are the major cause of soil, water and air pollution. Chemical fertilizers crowd out useful minerals naturally present in the topsoil. The microbes like bacteria, fungi, actinomycetes, worms etc. in top soil enrich the humus and help to produce nutrients to be taken up by the plants and later by animals. However, fertilizer enriched soil is unable to support microbial life and hence there is less humus and less nutrients and the soil easily becomes poor and eroded by rain and wind.

Excessive uses of nitrogenous fertilizer in modern farming decreases the potassium content of crops. Similarly excessive potash treatment decreases valuable nutrients in foods, such as ascorbic acid and carotene. The use of superphosphate leads to copper and zinc deficiency in crop plants. Nitrate fertilizer increase the crop yield (carbohydrate) but at the expense of proteins. Excessive fertilizer use produces over-sized fruits and vegetables, which are prone to insects and other pests.

The fertilizers used to raise the crop yield are drained by rain water to the adjacent fresh water bodies like rivers, lakes and ponds causing nutrient enrichment (especially nitrate and phosphate) of the aquatic bodies. This phenomenon is called as 'eutrophication', which triggers the luxuriant growth of blue green algae (cyanobacteria). The algal growth forms floating scums or blankets of algae called as algal blooms. Blooms of algae are generally not utilized by zooplanktons. The algal blooms compete for light for photosynthesis with other aquatic plants. Thus oxygen is depleted.

These blooms also release some toxic chemicals, which deteriorate the water quality. The decomposition of blooms also leads to oxygen depletion in water. Thus in poorly oxygenated water with higher carbon dioxide (CO_2) levels, fishes and other animals begin to die and clean water is turned into a stinking drain. The drinking of nitrate and nitrite contaminated water causes the disease 'methaemoglobinaemea' in children, which interferes with the oxygen carrying capacity of the blood. This leads to various disorders like damage to respiratory and vascular system, blue colouration of skin and even cancer.

The pesticides moving from crop fields to aquatic bodies affect the aquatic flora and fauna. Many of non-biodegradable pesticides (Chlorinated hydrocarbons) like D.D.T (Dichloro diphenyl trichloroethane), B.H.C (Benzene hexachloride) etc. enter the food chain and reach the human body causing harmful effect to human health.

The concentration of the pesticides increases with increasing food chain and the phenomenon is known as biological magnification. India's daily diet is reported to contain 270 µg of D.D.T and the level of accumulated D.D.T in the body tissue of an average Indian is highest in the world varying between 13 to 31.0 ppm (parts per million).

Cases of cancer, deformities, hepatic diseases and neurological disorders from pesticide poisoning have been reported from cotton growing areas of Gujarat, Maharashtra and Andhra Pradesh. Pesticide endosulphan used in aerial spray by Plantation Corporation of Kerala over the areas of Kasaragod and Rajapuram cashew plantation lead to severe cases of child blindness, physical retardation and cancer in these areas.

The excessive use of nitrogenous fertilizers causes acidification of soil, resulting in the loss of soil fertility. The indiscriminate use of pesticides to control pests kills several useful flora and fauna of the soil, which promote soil fertility. Besides the targeted insects, useful insects promoting cross and self-pollination are also killed. This leads to decline in crop productivity owing to reduced rate of pollination accomplished by insects.

Microbial action on nitrogenous fertilizer in the soil leads to formation of nitrous oxide, which causes the thinning of stratospheric ozone layer. The latter is a protective shield filtering harmful ultraviolet radiation emanating from the sun. Excessive use of water for irrigation in modern agriculture leads to water logging which causes anaerobic condition in the soil resulting in the production of methane gas by methanogenic bacteria. Methane (CH_4) is a potent green house gas causing global warming is increasing at the rate of 1% per annum.

In modern farming practices there is constant use of some high yielding varieties of the crops in place of nutritive indigenous varieties resulting into uniformity in biodiversity. This poses threat to the loss of biodiversity. The gradual loss of variability in the cultivated forms and in their wild relatives is referred to as 'genetic erosion'. This variability arose in nature over an extremely long period of time, and if lost would not be reproduced during a short time period. In modern farming, the loss of biological diversity is enhanced due to overexploitation of natural resources, excessive use of pesticides and environmental pollution.

The various aforesaid side effects of expensive modern farming on soil, crops and human health have compelled to look for an alternative in the form of organic farming. The latter is inexpensive, sound, safe and sustainable in long run without any adverse impact on the environment. Therefore, the organic farming has become inevitable to tackle with the problem of land degradation, environmental pollution, biodiversity depletion and contamination of food grains from pesticide residues.

Management Practices in Organic Farming

The management practices for organic farming differ from those of modern farming. The important steps in this type of farming are conservation of soil and genetic resources, integrated nutrient management, integrated weed management and integrated pest management.

Tillage practices in organic farming aim at reducing soil degradation. Therefore, conservational tillage is adopted in place of conventional tillage. Conservational tillage is disturbing the soil to the minimum extent necessary and leaving crop residues on the soil. Zero tillage and minimum tillage are the types of conservational tillage which reduce soil loss up to 99% over conventional tillage. In most cases, conservation tillage reduces soil loss by 50% over conventional tillage. Moreover, conservational tillage maintains the organic matter content of the soil and prevents the removal of nutrients from soil through rainwater. Conservational tillage also causes an increase in microbial and earthworms population in the soil.

Organic farming emphasizes on the cultivation of different indigenous nutritive local varieties of crops in place of few high yielding hybrid varieties only.

In organic farming, besides manure, green manure, compost and vermi-compost, oil cakes and oil meals play a key role as natural fertilizers. The commonly used organic nitrogenous fertilizers include rapeseed, mustard, neem, castor, mahua, karanja and linseed cakes. In addition to these, cakes from sal, groundnut and soyabean are also used in various combinations to increase yield and control pests.

Organic fertilizers have a slower action but they supply available nitrogen over a longer period of time. Besides, they protect useful flora and fauna of the soil; ameliorate yields and quality of products. Since there is increase in soil fertility, the biological activity is maintained intact.

In organic farming nitrogenous bio-fertilizers like Azolla pinnata (Pteridophyte), Anabaena, Aulosira, Nostoc, Scytonema, Tolypothrix, Cylinderospermum, Camptylonema, Westiellopsis (Blue green algae) and Azorhizobium, Bradyrhizobium, Mesorhizobium, Rhizobium, Sinorhizobium, Azotobacter Azospirillum and (Bacteria) are used to raise the fertility of soil. Furthermore, the fungi Aspergillus, Penicillium and Trichoderma are used as cellulolytic bio-fertilizers to enhance the rate of organic matter decomposition for the quick release of nutrients in the soil. The bacteria Bacillus subtilis, Pseudomonas putida and Pseudomonas fluorescens are used as phosphatic bio-fertilizers to solubilize phosphate.

The application of bio-fertilizers like Azospirillum makes the plant hardier by producing certain phenolic substances and eventually provides resistance to pest and diseases. Blue green algae not only fix atmospheric nitrogen but also excrete vitamin B12, ascorbic acid and auxins, which improve the growth of crop plants. They also possess the properties of solubilizing the bound phosphate of the soil.

In organic farming the weeds are controlled by employment of physical, cultural and biological method. Insect pests are controlled by combination of cultural and biological methods and at the same time use of resistant crop varieties.

Crop rotation practices are key to success of organic farming. Crop rotation is not only important for the soil fertility management but is also helpful in weed, insects and disease control. Legumes are essential in rotation practices for nitrogen supplement to the soil. The practice of mixed cropping increases the crop yield and avoids the chances of disease occurrence and pest infestation.

Principles of Organic Farming

Principle of Health

Organic Agriculture should sustain and enhance the health of soil, plant, animal, human and planet as one and indivisible.

This principle points out that the health of individuals and communities cannot be separated from the health of ecosystems - healthy soils produce healthy crops that foster the health of animals and people.

Health is the wholeness and integrity of living systems. It is not simply the absence of illness, but the maintenance of physical, mental, social and ecological well-being. Immunity, resilience and regeneration are key characteristics of health.

The role of organic agriculture, whether in farming, processing, distribution, or consumption, is to sustain and enhance the health of ecosystems and organisms from the smallest in the soil to human beings. In particular, organic agriculture is intended to produce high quality, nutritious food that contributes to preventive health care and well-being. In view of this it should avoid the use of fertilizers, pesticides, animal drugs and food additives that may have adverse health effects.

Principle of Ecology

Organic Agriculture should be based on living ecological systems and cycles, work with them, emulate them and help sustain them.

This principle roots organic agriculture within living ecological systems. It states that production is to be based on ecological processes, and recycling. Nourishment and well-being are achieved through the ecology of the specific production environment. For example, in the case of crops this is the living soil; for animals it is the farm ecosystem; for fish and marine organisms, the aquatic environment.

Organic farming, pastoral and wild harvest systems should fit the cycles and ecological balances in nature. These cycles are universal but their operation is site-specific. Organic management must be adapted to local conditions, ecology, culture and scale. Inputs should be reduced by reuse, recycling and efficient management of materials and energy in order to maintain and improve environmental quality and conserve resources.

Organic agriculture should attain ecological balance through the design of farming systems, establishment of habitats and maintenance of genetic and agricultural diversity. Those who produce, process, trade, or consume organic products should protect and benefit the common environment including landscapes, climate, habitats, biodiversity, air and water.

Principle of Fairness

Organic Agriculture should build on relationships that ensure fairness with regard to the common environment and life opportunities.

Fairness is characterized by equity, respect, justice and stewardship of the shared world, both among people and in their relations to other living beings.

This principle emphasizes that those involved in organic agriculture should conduct human relationships in a manner that ensures fairness at all levels and to all parties - farmers, workers, processors, distributors, traders and consumers. Organic agriculture should provide everyone involved with a good quality of life, and contribute to food sovereignty and reduction of poverty. It aims to produce a sufficient supply of good quality food and other products.

This principle insists that animals should be provided with the conditions and opportunities of life that accord with their physiology, natural behavior and well-being.

Natural and environmental resources that are used for production and consumption should be managed in a way that is socially and ecologically just and should be held in trust for future generations. Fairness requires systems of production, distribution and trade that are open and equitable and account for real environmental and social costs.

Principle of Care

Organic Agriculture should be managed in a precautionary and responsible manner to protect the health and well-being of current and future generations and the environment.

Organic agriculture is a living and dynamic system that responds to internal and external demands and conditions. Practitioners of organic agriculture can enhance efficiency and increase productivity, but this should not be at the risk of jeopardizing health and well-being. Consequently, new technologies need to be assessed and existing methods reviewed. Given the incomplete understanding of ecosystems and agriculture, care must be taken.

This principle states that precaution and responsibility are the key concerns in management, development and technology choices in organic agriculture. Science is necessary to ensure that organic agriculture is healthy, safe and ecologically sound. However, scientific knowledge alone is not sufficient. Practical experience, accumulated wisdom and traditional and indigenous knowledge offer valid solutions, tested by time. Organic agriculture should prevent significant risks by adopting appropriate technologies and rejecting unpredictable ones, such as genetic engineering. Decisions should reflect the values and needs of all who might be affected, through transparent and participatory processes.

Organic Farming and Biodiversity

The effect of organic farming has been a subject of interest for researchers. Theory suggests that organic farming practices, which exclude the use of most synthetic pesticides and fertilizers, may be beneficial for biodiversity. This is generally shown to be true for soils scaled to the area of cultivated land, where species abundance is, on average, 30% richer than that of conventional farms. However, for crop yield-scaled land the effect of organic farming on biodiversity is highly debated due to the significantly lower yields compared to conventional farms.

In ancient farming practices, farmers did not possess the technology or manpower to have a significant impact on the destruction of biodiversity even as mass-production agriculture was rising. Nowadays, common farming methods generally rely on pesticides to maintain high yields. With such, most agricultural landscapes favor mono-culture crops with very little flora or fauna co-existence. Modern organic farm practices such as the removal of pesticides and the inclusion of animal manure, crop rotation, and multi-cultural crops provides the chance for biodiversity to thrive.

Benefits of Organic Farming to Biodiversity

Nearly all non-crop, naturally occurring species observed in comparative farm land practice studies show a preference in organic farming both by population and richness. Spanning all associated species, there is an average of 30% more on organic farms versus conventional farming methods, however this does not account for possible loss of biodiversity due to decreased yields. Birds, butterflies, soil microbes, beetles, earthworms, spiders, vegetation, and mammals are particularly affected. Some organic farms may use less pesticides and thus biodiversity fitness and population density may benefit. Larger farms however tend to use pesticides more liberally and in some cases to larger extent than conventional farms. Many weed species attract beneficial insects that improve soil qualities and forage on weed pests. Soil-bound organisms often benefit because of increased bacteria populations due to natural fertilizer spread such as manure, while experiencing reduced intake of herbicides and pesticides commonly associated with conventional farming methods. Increased biodiversity, especially from soil microbes such as mycorhizzae, have been proposed as an explanation for the high yields experienced by some organic plots, especially in light of the differences seen in a 21-year comparison of organic and control fields.

Impact of Increased Biodiversity

The level of biodiversity that can be yielded from organic farming provides a natural capital to humans. Species found in most organic farms provides a means of agricultural sustainability by reducing amount of human input (e.g. fertilizers, pesticides). Farmers that produce with organic methods reduce risk of poor yields by promoting biodiversity. Common game birds such as the ring-necked pheasant and the northern bobwhite often reside in agriculture landscapes, and are a natural capital yielded from high demands of recreational hunting. Because bird species richness and population are typically higher on organic farm systems, promoting biodiversity can be seen as logical and economical.

Highly Impacted Animal Species

Earthworms

Earthworm population and diversity appears to have the most significant data out of all studies. Out of six studies comparing earthworm biodiversity to organic and conventional farming methods, all six suggested a preference for organic practices including a study at the pioneering Haughley farm in 1980/1981 that compared earthworm populations and soil properties after 40 years. Hole et al. summarized a study conducted by Brown and found nearly double the population and diversity when comparing farming methods.

Birds

Organic farms are said to be beneficial to birds while remaining economical. Bird species are one of the most prominent animal groups that benefit from organic farming methods. Many species rely on farmland for foraging, feeding, and migration phases. With such, bird populations often relate directly to the natural quality of farmland. The more natural diversity of organic farms provides better habitats to bird species, and is especially beneficial when the farmland is located within a migration zone. In 5 recent studies almost all bird species including locally declining species, both population and variation increased on organic farmland,. Making a switch from conventional farming methods to organic practices also seems to directly improve bird species in the area. While organic farming improves bird populations and diversity, species populations receive the largest boost when organic groups are varied within a landscape. Bird populations are increased further with optimal habitat for biodiversity, rather than organic alone, with systems such as Conservation Grade.

Butterflies

A specific study done in the UK in 2006 found substantially more butterflies on organic farms versus standard farming methods except for two pest species. The study also observed higher populations in uncropped field margins compared with cropland edges regardless of farm practice. Conversely, Weibull et al. found no significant differences in species diversity or population.

Spiders

Ten studies have been conducted involving spider species and abundance on farm systems. All but three of the studies indicated that there was a higher diversity of spider species on organic farms, in addition to populations of species. Two of the studies indicated higher species diversity, but statistically insignificant populations between organic and standard farming methods.

Soil Microbes

Out of 13 studies comparing bacteria and fungus communities between organic and standard farming, 8 of the studies showed heightened level of growth on organic farm systems. One study concluded that the use of "green" fertilizers and manures was the primary cause of higher bacterial levels on organic farms. On the other hand, nematode population/diversity depended on what their primary food intake was. Bacteria-feeding nematodes showed preference towards organic systems whereas fungus-feeding nematodes showed preference for standard farm systems. The heightened level of bacteria-feeding nematodes makes sense due to higher levels of bacteria in organic soils, but the fungus-feeding populations being higher on standard farms seems to contradict the data since more fungi are generally found on organic farms.

Beetles

According to Hole et al. beetle species are among the most commonly studied animal species on farming systems. Twelve studies have found a higher population and species richness of carabids on organic systems. The overall conclusion of significantly higher carabid population species and diversity is that organic farms have a higher level of weed species where they can thrive. Staphylinid populations and diversity have seemed to show no specific preference with some studies showing higher population and diversity, some with lower population and diversity, and one study showed no statistical significance between the organic and conventional farming systems.

Mammals

Two comparative studies have been conducted involving mammal populations and diversity among farm practices. A study done by Brown found that small mammal population density and diversity did not depend on farming practices, however overall activity was higher on organic farms. It was concluded that more food resources were available to small mammals on organic farms because of the reduction or lack of herbicides and pesticides. Another study conducted by Wickramasinghe et al. compared bat species and activity. Species activity and foraging were both more than double on organic farms compared to conventional farms. Species richness was also higher on organic farms, and 2 of the sixteen species sighted were found only on organic farms.

Vegetation

Approximately, ten studies have been conducted to compare non-crop vegetation between organic and conventional farming practices. Hedgerow, inner-crop and grassland observations were made within these studies and all but one showed a higher weed preference and diversity in or around organic farms. Most of these studies showed significant overall preference for organic farming preferences especially for broad-leafed species, but many grass species showed far less on conventional farms likely because pesticide interaction was low or non-existent. Organic farm weed population and richness was believed to be lower in mid-crop land because of weed-removal methods such as under sowing. Switching from conventional to organic farming often results in a "boom" of weed speciation due to intense chemical change of soil composition from the lack of herbicides and pesticides. Natural plant species can also vary on organic farms from year-to-year because crop rotation creates new competition based on the chemical needs of each crop.

Farmers' Benefits from Increased Biodiversity

Biological research on soil and soil organisms has proven beneficial to the system of organic farming. Varieties of bacteria and fungi break down chemicals, plant matter and animal waste into productive soil nutrients. In turn, the producer benefits by healthier yields and more arable soil for future crops. Furthermore, a 21-year study was conducted testing the effects of organic soil matter and its relationship to soil quality and yield. Controls included actively managed soil with varying levels of manure, compared to a plot with no manure input. After the study commenced, there was significantly lower yields on the control plot when compared to the fields with manure. The concluded reason was an increased soil microbe community in the manure fields, providing a healthier, more arable soil system.

Detriments to Biodiversity through Organic Farming

Organic farming practices still require active participation from the farmer to effectively boost biodiversity. Making a switch to organic farming methods does not automatically or guarantee improved biodiversity. Pro-conservation ethics are required to create arable farm land that generates biodiversity. Conservationist ideals are commonly overlooked because they require additional physical and economical efforts from the producer. Common weed-removal processes like undercutting and controlled burning provides little opportunity for species survival, and often leads to comparable populations and richness to conventionally managed landscapes when performed in excess. Another common process is the addition of biotopes in the form of hedgerows and ponds to further improve species richness. Farmers commonly make the mistake of over-using these resources for more intense crop production because organic yields are typically lower. Another error comes from the over-stratification of biotopes. A series of small clusters does not provide adequate land area for high biodiversity potential.

Organic Horticulture

An organic garden on a school campus.

Organic horticulture is the science and art of growing fruits, vegetables, flowers, or ornamental plants by following the essential principles of organic agriculture in soil building and conservation, pest management, and heirloom variety preservation.

Horticulture is also sometimes defined simply as "agriculture minus the plough." Instead of the plough, horticulture makes use of human labour and gardener's hand tools, although some small machine tools like rotary tillersare commonly employed now.

General

Mulches, cover crops, compost, manures, vermicompost, and mineral supplements are soil-building mainstays that distinguish this type of farming from its commercial counterpart. Through attention to good healthy soil condition, it is expected that insect, fungal, or other problems that sometimes plague plants can be minimized. However, pheromone traps, insecticidal soap sprays, and other pest-control methods available to organic farmers are also utilized by organic horticulturists.

Horticulture involves five areas of study. These areas are floriculture (includes production and marketing of floral crops), landscape horticulture (includes production, marketing and maintenance of landscape plants), olericulture (includes production and marketing of vegetables), pomology (includes production and marketing of fruits), and postharvest physiology (involves maintaining quality and preventing spoilage of horticultural crops). All of these can be, and sometimes are, pursued according to the principles of organic cultivation.

Organic horticulture (or organic gardening) is based on knowledge and techniques gathered over thousands of years. In general terms, organic horticulture involves natural processes, often taking place over extended periods of time, and a sustainable, holistic approach - while chemical-based horticulture focuses on immediate, isolated effects and reductionist strategies.

Organic Gardening Systems

There are a number of formal organic gardening and farming systems that prescribe specific techniques. They tend to be more specific than, and fit within, general organic standards. Forest gardening, a fully organic food production system which dates from prehistoric times, is thought to be the world's oldest and most resilient agroecosystem.

Biodynamic farming is an approach based on the esoteric teachings of Rudolf Steiner. The Japanese farmer and writer Masanobu Fukuoka invented a no-till system for small-scale grain production that he called Natural Farming. French intensive gardening and biointensive methods and SPIN Farming (Small Plot INtensive) are all small scale gardening techniques. These techniques were brought to the United States by Alan Chadwick in the 1930s. A garden is more than just a means of providing food, it is a model of what is possible in a community - everyone could have a garden of some kind (container, growing box, raised bed) and produce healthy, nutritious organic food, a farmers market, a place to pass on gardening experience, and a sharing of bounty, promoting a more sustainable way of living that would encourage their local economy. A simple 4' x 8' (32 square feet) raised bed garden based on the principles of bio-intensive planting and square foot gardening uses fewer nutrients and less water, and could keep a family, or community, supplied with an abundance of healthy, nutritious organic greens, while promoting a more sustainable way of living.

Organic gardening is designed to work with the ecological systems and minimally disturb the Earth's natural balance. Because of this organic farmers have been interested in reduced-tillage methods. Conventional agriculture uses mechanical tillage, which is plowing or sowing, which is harmful to the environment. The impact of tilling in organic farming is much less of an issue. Ploughing speeds up erosion because the soil remains uncovered for a long period of time and if it has a low content of organic matter, the structural stability of the soil decreases. Organic farmers use techniques such as mulching, planting cover crops, and intercropping, to maintain a soil cover throughout most of the year. The use of compost, manure mulch and other organic fertilizers yields a higher organic content of soils on organic farms and helps limit soil degradationand erosion.

Other methods such as composting or vermicomposting can also be used to supplement an existing garden. These practices are ways of recycling organic matter into some of the best organic fertilizers and soil conditioner. Vermicompost is especially easy. The byproduct is also an excellent source of nutrients for an organic garden.

Pest Control Approaches

Differing approaches to pest control are equally notable. In chemical horticulture, a specific insecticide may be applied to quickly kill off a particular insect pest. Chemical controls can dramatically reduce pest populations in the short term, yet by unavoidably killing (or starving) natural control insects and animals, cause an increase in the pest population in the long term, thereby creating an ever-increasing problem. Repeated use of insecticides and herbicides also encourages rapid natural selection of resistant insects, plants and other organisms, necessitating increased use, or requiring new, more powerful controls.

In contrast, organic horticulture tends to tolerate some pest populations while taking the long view. Organic pest control requires a thorough understanding of pest life cycles and interactions, and involves the cumulative effect of many techniques, including:

- Allowing for an acceptable level of pest damage.

- Encouraging predatory beneficial insects to flourish and eat pests.

- Encouraging beneficial microorganisms.

- Careful plant selection, choosing disease-resistant varieties.

- Planting companion crops that discourage or divert pests.

- Using row covers to protect crop plants during pest migration periods.

- Rotating crops to different locations from year to year to interrupt pest reproduction cycles.

- Using insect traps to monitor and control insect populations.

Each of these techniques also provides other benefits, such as soil protection and improvement, fertilization, pollination, water conservation and season extension. These benefits are both complementary and cumulative in overall effect on site health. Organic pest control and biological pest control can be used as part of integrated pest management (IPM). However, IPM can include the use of chemical pesticides that are not part of organic or biological techniques.

Impact on the Global Food Supply

One controversy associated with organic food production is the matter of the amount of food produced per acre. Even with good organic practices, organic agriculture may be five to twenty-five percent less productive than conventional agriculture, depending on the crop.

Much of the productivity advantage of conventional agriculture is associated with the use of nitrogen fertilizer. However, the use, and especially the overuse, of nitrogen fertilizer has negative effects such as nitrogen runoff harming natural water supplies and increased global warming.

Organic methods have other advantages, such as healthier soil, that may make organic farming more resilient, and therefore more reliable in producing food, in the face of challenges such as climate change.

As well, world hunger is not primarily an issue of agricultural yields, but distribution and waste.

Organic Aquaculture

Organic aquaculture is a holistic method for farming marine species in line with organic principles. The ideals of this practice established sustainable marine environments with consideration for naturally occurring ecosystems, use of pesticides, and the treatment of aquatic life. Managing aquaculture organically has become more popular since consumers are concerned about the harmful impacts of aquaculture on themselves and the environment.

The availability of certified organic aquaculture products have become more widely available since the mid-1990s. This seafood growing method has become popular in Germany, the United Kingdom and Switzerland, but consumers can be confused or skeptical about the label due to conflicting and misleading standards around the world.

A certified organic product seal on aquaculture products will mean an accredited certifying body has verified that the production methods meet or exceed a country's standard for organic aquaculture production. Organic regulations designed around soil-based systems don't transfer well into aquaculture and tend to conflict with large-scale, intensive (economically viable) practices/goals. There are a number of problems facing organic aquaculture: difficulty of sourcing and certifying organic juveniles (hatchery or sustainable wild stock); 35-40% higher feed cost; more labour-intensive; time and cost of the certification process; a higher risk of diseases, and uncertain benefits. But, there is a definite consumer demand for organic seafood, and organic aquaculture may become a significant management option with continued research.

Certification

A number of countries have created their own national standards and certifying bodies for organic aquaculture. While there is not simply one international organic aquaculture standardization process, one of the largest certification organizations is the Global Trust, which delivers assessments and certifications to match the highest quality organic aquaculture standards. The information regarding these standards is available through a personal inquiry.

Many organic aquaculture certifications address a variety of issues including antibiotic and chemical treatments of fish, unrestrained disposal of fish feces into the ocean, fish feeding materials, the habitat of where and how the fish are raised, and proper handling practices including slaughter. Most Organic Aquaculture certifications follow rather strict requirements and standards. These rules may vary between different countries or certification bodies. This leads to confusion when products are imported from other countries, which can result in a backlash from consumers (for example, the Pure Salmon Campaign).

Defining acceptable practices is also complicated by the variety of species - freshwater, saltwater, shellfish, finfish, mollusks and aquatic plants. The difficulty of screening pollutants out of an aquatic medium, controlling the food supplies and of keeping track of individual fish may mean that fish and shellfish stocks should not be classified as 'livestock' at all under regulations. This point further exemplifies the need for widespread aquaculture certification standard.

Challenges and Controversy

There is some controversy over licensing restrictions, as some seafood companies propose that wild caught fish should be classified as organic. While wild fish may be free of pesticides and unsustainable rearing practices, the fishing industry may not necessarily be environmentally sustainable.

The variation in standards, as well as the unknown level of actual compliance and the closeness of investigations when certifying are major problems in consistent organic certification. In 2010, new rules were proposed in the European Union to consistently define the organic aquaculture industry. Canada's General Standards Board's (CGSB) proposed updates to their standards were strongly opposed in 2010 because they allowed antibiotic and chemical treatments of fish, up to 30 percent non-organic feed, deadly and uncontrolled impacts on wild species and unrestrained disposal of fish feces into the ocean. These standards would have certified net pen systems as organic. At the other end of the scale, the extremely strict national legislation in Denmark has made it difficult for the existing organic trout industry to develop.

Potential alternatives to Non-organic Feed and Waste Removal

One major issue in organic aquaculture production is finding practical and sustainable alternatives to non-organic veterinary treatments, feeds, spat and waste disposal. Potential veterinary alternatives include homeopathic treatments and production-cycle limited allopathic or chemical treatments. Current requirements usually stipulate a reduction in unsustainable fishmeal, in favor of organic vegetable and fish by-product replacements. A recent study into organic fish feeds for salmon found that while organic feed provide some benefit to the environmental impact of the fishes' life cycles, the loss of fish meals and oils have a significant negative impact. Another study discovered that certain percentages of dietary protein could be safely replaced.

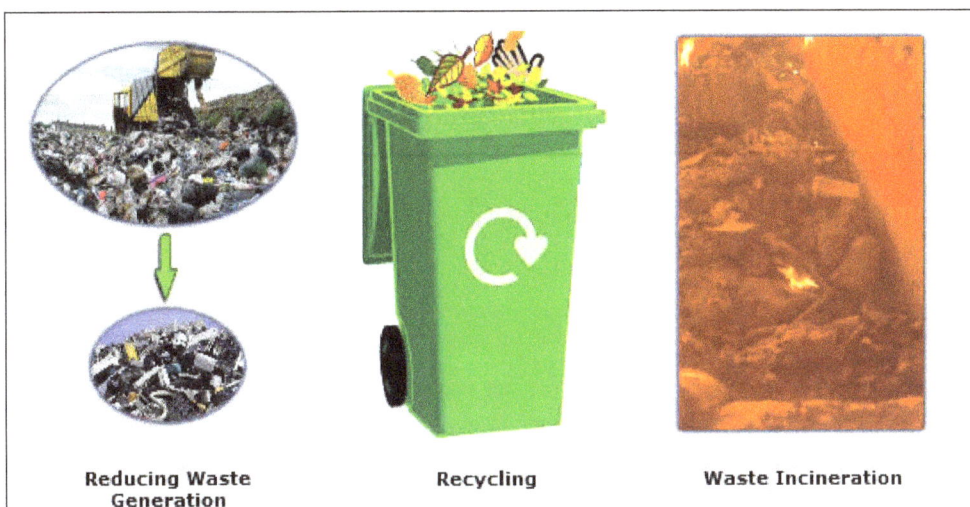

Reducing Waste Generation   Recycling   Waste Incineration

Not only do the fish have to be organically reared, organic fish feeds need to be developed. Research into ways of decreasing the amount on non-sustainable fishmeal in feed is currently focusing on replacement by organic vegetable proteins. Some organic fish feeds becoming available, and/or the option of integrated multi-species systems (e.g. growing plants using aquaponics, as well as larvae or other fish). For example, locating a shellfish bed next to a finfish farm to dispose of the waste and provide the shellfish with controlled nutrients.

United States Organic Aquaculture Certification

In 2005, with the growing need for a certification process specifically designed for marine-based farming methods, the National Organic Standards Board and the National Organics Program created a working group called the Aquatic Animal Task Force in order to seek recommendations for the new certification process. The task force was meant to be broken into two divisions: wild fisheries and aquaculture, but the wild fisheries group never materialized.

In 2006, the Aquaculture Working Group delivered a report with suggestions for the production and handling of aquatic animals and plants. However, with the complexity and diversity of the marine systems, the group requested more time to explore bivalve mollusks (oysters, clams, mussels and scallops) in depth. The National Organic Standards Board approved the aquaculture standards in 2007 and reconsidered the aquatic animal feed and facilities until they synthesized the public commentary in 2008. In 2010, the NOSB approved the recommendations for the bivalve mollusks section.

Currently, the legal status of using the organic label for aquatic species, and the future of developing U.S. Department of Agriculture (USDA) certification standards for organic aquaculture products and aquatic species, are under review. It is anticipated that the first version of the rule for organic aquaculture will be announced in April or May 2016 with need for approval by the Office of Management and Budget. It is expected to see the final rule in play by late summer or fall of 2016 with organic aquaculture products likely available in store in 2017. The certification is said to include the following: shellfish, marine and recirculating system methods of aquaculture, as well as the controversial net-pen method.

The US currently allows the imports of organically-certified seafood from Europe, Canada and other countries around the world.

Production

Organic aquaculture was responsible for an estimated US$46.1 billion internationally (2007). There were 0.4 million hectares of certified organic aquaculture in 2008 compared to 32.2 million hectares dedicated to Organic farming. The 2007 production was still only 0.1% of total aquaculture production.

The market for organic aquaculture shows strong growth in Europe, especially France, Germany and the UK - for example, the market in France grew 220% from 2007 to 2008. There is a preference for organic food, where available. Organic seafood is now sold in discount supermarket chains throughout the EU. The top five producing countries are UK, Ireland, Hungary, Greece and France. 123 of the 225 global certified organic aquaculture farms operate in Europe and were responsible for 50,000 tonnes in 2008 (nearly half global production).

Organic seafood products are a niche market and users currently expect to pay premiums of 30-40%. Organic salmon is the top species and retails at 50%. Market demand is driving Danish rainbow trout farmers to switch to organic farming.

Known Data on Organic Aquaculture by Country

Asia

Country	Organically managed area [ha]
Bangladesh	2'000

China	415'000
Ecuador	6'382
Indonesia	1'317
Thailand	33
Total	424'732

Indonesian Shrimp farms are locally certified as organic but a recent study found them to be highly environmentally damaging.

Europe

- Denmark: Rainbow Trout. Organic production ~400 tonnes (1% of total trout production).

- UK: Cod and carp, Trout, Salmon.

- Rainbow Trout (Denmark).

- Salmon (80% of organic aquaculture production in 2000) and shrimp (Europe).

- Carp (low volume production, poorly marketed - Europe).

North America

- Shellfish: oyster, clam, mussel, scallop, geoduck seed (USA).

Organic production of crops and livestock in the United States is regulated by the Department of Agriculture's National Organic Program (NOP). While it does cover aquaponics, it did not properly cover aquaculture until the recent 2008 amendment, hampering the progress of organic aquaculture in the states.

Australia

New Zealand

The first certified organic aquaculture farm in New Zealand was a salmon farm which was the largest producer outside of Europe contributing to the European market. New Zealand green-lipped mussel Greenshell mussels - certified by Sealord (12), DOM ORGANICS Greenshell mussels, certified organic by Bio-Gro New Zealand Ltd. (BGNZ).

Salmon (14) 12 tonnes/year - Ormond Aquaculture Ltd. certified (CERTNZ) organic freshwater aquaculture farm.

Koura (freshwater crayfish) Still being developed - Ormond Aquaculture Ltd certified (CERTNZ) organic freshwater aquaculture farm.

Biointensive Agriculture

Biointensive agriculture is an organic agricultural system that focuses on achieving maximum yields from a minimum area of land, while simultaneously increasing biodiversity and sustaining the fertility of the soil. The goal of the method is long term sustainability on a closed system basis. It is particularly effective for backyard gardeners and smallholder farmers in developing countries, and also has been used successfully on small-scale commercial farms.

Many of the techniques that contribute to the biointensive method were present in the agriculture of the ancient Chinese, Greeks, Mayans, and of the Early Modern period in Europe, as well as in West Africa (Tapades of Fouta Djallon) from at least the late 18th century. Alan Chadwick brought together the biodynamic and French intensive gardening methods, as well as his own unique approach, to form what he called the Biodynamic-French Intensive method.

The method was further developed by John Jeavons and Ecology Action into a sustainable 8-step food-raising method officially known as "grow biointensive sustainable mini-farming". The method now enjoys widespread practice and further development, and according to Ecology Action, has been used in over 140 countries around the world, in almost every climate and soil where food is grown. Components important to the biointensive approach include:

- Double-dug and raised beds,

- Composting,

- Biointensive planting,

- Companion planting,

- Carbon farming,

- Calorie farming,

- Use of open-pollinated seeds,

- A whole-system farming method.

But that concept and method have dealt with only eco-technical aspect. Rajbhandari further developed the holistic concept and approach of bio-intensive farming system to address the socio-economic, cultural and political aspects as well. Rajbhandari has defined bio-intensive farming system (BIFS) as a biologically intensive mixed farming system, which relies on the intensive engagement of the farmers; optimization of organic recycling through crop rotations; integrated plant nutrient management (IPNM); and integrated organic pest management (IOPM) with the use of bio-pesticides, botanical pesticides, and biota e.g. Trichogramma chilonis. The IPNM in BIFS is provisioned to include improved FYM, compost, green manure and bio-fertilizers (azola, Rhizobium and Mycorrhizal). It is a holistic system of sustainable management of natural resources in a given agro-ecosystem with specific cultural and knowledge base.

Sustainable bio-intensive farming (BIF) system, which emphasizes biodiversity conservation; recycling of nutrients; synergy among crops, animals, soils, and other biological components; and regeneration and conservation of resources is a type of agro-ecological approach. This is the alternative approach that can appropriately address the central issue of hunger, poverty, food/nutrition insecurity and livelihoods. It has been serving as a model for promoting ecological farms and eco-tourism for higher productions and income generation in small scale.

System

The biointensive method provides many benefits as compared with conventional farming and gardening methods, and is an inexpensive, easily implemented sustainable production method that can be used by people who lack the resources (or desire) to implement commercial chemical and fossil-fuel-based forms of agriculture.

Ecology Action's research shows that biointensive methods can enable small-scale farms and farmers to significantly increase food production and income, utilize predominantly local, renewable resources and decrease expense and energy inputs while building fertile topsoil at a rate 60 times faster than in nature.

According to Jeavons and other proponents, when properly implemented, farmers using biointensive techniques have the potential to:

- Use 67% to 88% less water than conventional agricultural methods.

- Use 50% to 100% less purchased (organic, locally available) fertilizer.

- Use up to 99% less energy than commercial agriculture, while using a fraction of the resources.

- Produce 2 to 6 times more food at intermediate yields, assuming a reasonable level of farmer skill and soil fertility (which increase over time as the method is practiced).

- Produce a 100% increase in soil fertility.

- Reduce by 50% or more the amount of land required to grow a comparable amount of food. This allows more land to remain in a wild state, preserving ecosystem services and promoting genetic diversity.

In order to achieve these benefits, the biointensive method uses an eight-part integrated system of deep soil cultivation ("double-digging") to create raised, aerated beds;

intensive planting; companion planting; composting; the use of open-pollinated seeds; and a carefully balanced planting ratio of 60% Carbon-Rich Crops (for compost production) 30% Calorie-Rich Crops (for food) and an optional 10% planted in Income Crops (for sale).

The following outline of the methods approximates the descriptions found in the popular biointensive handbook, How to Grow More Vegetables and fruits, nuts, berries, grains and other crops. Than You Ever Thought Possible on Less Land Than You Can Imagine, by John Jeavons, now in its eighth edition, and in seven languages, including braille.

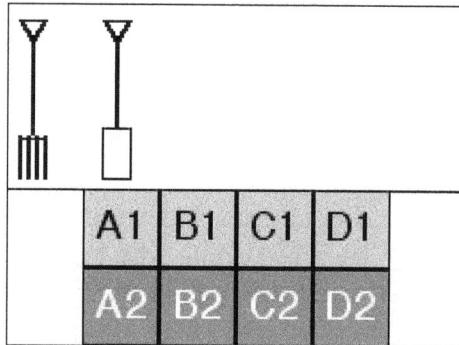

- In double digging, a 12-inch (305 mm) deep trench is dug across the width of the bed with a flat spade, and the soil from that first trench is set aside. The 12 inches (305 mm) below the trench are loosened with a spading fork. When the next trench is dug, that soil is dropped into the empty space of the first trench, and the lower layer is again loosened with a spading fork. This process is repeated along the full length of the bed. The final trench is filled with the soil that was removed from the first trench. The result is a bed that has been tilled to a depth of 24 inches (610 mm). When an entire bed has been double dug, the soil will have greater drainage and aeration, which allows the roots to grow much deeper and reach more nutrients. Despite the fact that no soil has been added, the bed is raised due to the aeration. It is worth noting that hard, unworked soil should be double dug each season until the soil has attained good structure and long lasting aeration. During subsequent seasons, it can be surface cultivated 2 to 4 inches (5 to 10 cm) deep with a hula hoe until compaction again becomes apparent. After double digging the first season, deep tilling during subsequent seasons can be quickly accomplished with a u-bar, particularly in the cases of larger minifarms or commercial farms.

- Composting allows the plants to transform and enrich the soil with organic matter, and also to return nutrients to the soil. Biointensive composting is fairly straightforward, emphasizing the health and diversity of the microbes that break down and become a part of the compost. Thus, relatively cooler composting is practiced, and plant materials are preferred over animal materials.

Soil is often combined with the compost to inoculate the pile with microbes. Without human waste recycling, however, nutrients and organic matter are constantly removed from the soil (as food that is consumed by the farmer) and flushed away. Therefore, when safe and legal human waste recycling is possible—as in many places it already is—that fertility can, and should, be returned to the soil. Another great unappreciated source of compost and soil improvement is the roots of crops themselves, which, in the biointensive system are left to decompose in the soil, where they help to both fertilize and "sew it together", creating stable soil structure. Thus, crops such as alfalfa, which has exceptionally deep roots, and cereal rye, which has a particularly high volume of roots, are valued.

- The soil air from the development of deep soil structure, combined with the microbe-and nutrient-rich compost allow the crops to be planted intensively. To plant intensively, beds are 4 to 6 feet (1.2 to 1.8 m) wide, usually 5 ft (1.5 m) and at least 5 feet (1.5 m) long, often 20 feet (6 m), forming a bed of 100 square feet (10 m^2). Crops are not planted in traditional rows according to a square pattern, but are planted in a hexagonal or triangular pattern in the bed so that no space is left unnecessarily unused. These wide beds and close spacings not only allow more plants per area (up to 4 times as many), but also enable the plants to form a living mulch over the soil, keeping in moisture and shading out weeds. Additionally, whenever possible seedlings are started in flats or nursery beds, so that more garden space is available to large plants and so that the seedlings can be more closely spaced before transplant, forming a living mulch in the flat as well.

- Companion planting is described as taking place both in space, which is traditionally called companion planting, and in time, which is traditionally called crop rotation. Companion planting can be used to improve the health and growth of crops, and also as another form of intensive planting, which uses vertical space more efficiently by mixing shallow rooting plants with deep rooting plants or slow growing plants with fast growing plants.

- In order to achieve sustainable fertility on a closed system basis, the biointensive method uses carbon and calorie farming, an aikido-style of work (using the least amount of energy or effort to achieve the greatest amount of work or production), composting—including safe and legal human waste recycling—the use of open pollinated seeds, and limited land use, which allows farmers and gardeners to retain more of the land in a wild state for genetic diversity and an ecosystem balance.

- If carbon or compost crops are grown in about sixty percent of the cultivated land, they can provide the compost materials that maintain the fertility for one hundred percent of the cultivated land. Many cereal crops qualify as compost crops, but provide both food and abundant compost. Some of the compost crops may be grown during the winter, when the land would be otherwise unused. Certain compost crops are higher in carbon while others are higher in nitrogenand/or fix nitrogen in the soil, and the desired proportion of each must be grown for the compost to achieve maximum effectiveness. Also, certain compost crops take particular desired nutrients from the subsoil and concentrate them in the compost, thus allowing a redistribution of those nutrients to the food crops. This proportion of 60% compost crops is crucial to the sustainability that is the goal of the biointensive method, and to the fertility of the garden.

• In calorie farming, care is given to growing enough food energy (and other nutrients) to live on in a minimal area. Root crops are often used in calorie farming because they allow biointensive farmers and gardeners to grow more nutrients in smaller areas, resulting in less labor per calorie, and more space for wilderness and other people. These crops—which have both a high calorie content per pound, and a high yield per area—include potatoes, sweet potatoes, garlic, leeks, burdock, Jerusalem artichoke and parsnips. These crops can produce as much as 5 to 20 times the calories per unit of area per unit of time. In biointensive farming, 30% of the land cultivated for food is used for root crops.

• The use of open pollinated seeds ensures genetic diversity, and allows the farmer to be self-sufficient, harvesting seeds from his or her own plants, and cultivating varieties which are best suited to that particular region.

• The Whole System: biointensive experts emphasize that because these techniques can result in intense productivity and high yields, the system must be practiced as a whole in order to prevent rapid soil exhaustion. The goal of the biointensive method is sustainability, but if the techniques concerning productivity are practiced without integrating the techniques concerning sustainable soil fertility, the soil may be depleted even more rapidly than with conventional farming methods. The most important element for building and maintaining sustainable soil fertility is the growing of 60% compost crops, proper composting techniques that incorporate the right balance of mature carbonaceous brown and green nitrogenous compost materials, and when possible, safe and legal human waste recycling.

Animals

The biointensive method typically concentrates on the vegan diet. This does not mean that biointensive farming must exclude the raising of animals. Animals, while not considered by biointensive practitioners to be sustainable, can be incorporated into biointensive systems, although they increase the amount of land and labor required considerably.

"Livestock can fit into a (biointensive) system, but it usually takes a larger area (than growing a vegan diet). Normally it takes about 40,000 sq ft of grazing land for 1 cow/steer (for milk/meat) or 2 goats (for milk/meat/wool), or 2 sheep (for milk/meat/wool). In contrast with biointensive farming and maximizing the edible calorie output in your vegan diet design, one person's complete balanced diet can be grown on about 4,000 sq ft—a much smaller area."

The challenge to growing animals for food is that by 2014, 90% of the world's people will only have access to about 4,500 sq ft of farmable land per person, if they leave an equal area in a wild state to protect plant and animal genetic diversity and the world's ecosystems. As you will see from the information that follows on the land requirements for incorporating livestock, this becomes a challenge.

The topic goes on to estimate the square footage required to grow fodder for various animals (and compost to replenish the soil), and provides a discussion on whether animal manure should be used as a fertilizer/compost supplement.

Organic Food

Organic produce at a farmers' market.

Organic food is food produced by methods that comply with the standards of organic farming. Standards vary worldwide, but organic farming features practices that cycle resources, promote ecological balance, and conserve biodiversity. Organizations regulating organic products may restrict the use of certain pesticidesand fertilizers in the farming methods used to produce such products. Organic foods typically are not processed using irradiation, industrial solvents, or synthetic food additives.

In the 21st century, the European Union, the United States, Canada, Mexico, Japan, and many other countries require producers to obtain special certification to market their food as organic. Although the produce of kitchen gardens may actually be organic, selling food with an organic label is regulated by governmental food safety-authorities, such as the National Organic Program of the US Department of Agriculture (USDA) or European Commission (EC).

From an environmental perspective, fertilizing, overproduction, and the use of pesticides in conventional farming may negatively affect ecosystems, biodiversity, groundwater, and drinking water supplies. These environmental and health issues are intended to be minimized or avoided in organic farming. However, the outcome of farming organically may not produce such benefits because organic agriculture has higher production costs and lower yields, higher labor costs, and higher consumer prices.

Demand for organic foods is primarily driven by consumer concerns for personal health and the environment. From the perspective of science and consumers, there is insufficient evidence in the scientific and medical literature to support claims that organic food is either safer or healthier to eat than conventional food. While there may be some differences in the nutrient and antinutrient contents of organically and conventionally produced food, the variable nature of food production, shipping, storage, and handling makes it difficult to generalize results. Claims that "organic food tastes better" are generally not supported by tests.

For the vast majority of its history, agriculture can be described as having been organic; only during the 20th century was a large supply of new products, generally deemed not organic, introduced into food production. The organic farming movement arose in the 1940s in response to the industrialization of agriculture.

In 1939, Lord Northbourne coined the term *organic farming*, out of his conception of "the farm as organism," to describe a holistic, ecologically balanced approach to farming—in contrast to what he called *chemical farming,* which relied on "imported fertility" and "cannot be self-sufficient nor an organic whole." Early soil scientists also described the differences in soil composition when animal manures were used as "organic", because they contain carbon compounds where superphosphates and haber process nitrogen do not. Their respective use affects humus content of soil. This is different from the scientific use of the term "organic" in chemistry, which refers to a class of molecules that contain carbon, especially those involved in the chemistry of life. This class of molecules includes everything likely to be considered edible, and include most pesticides and toxins too, therefore the term "organic" and, especially, the term "inorganic" as they apply to organic chemistry is an equivocation fallacy when applied to farming, the production of food, and to foodstuffs themselves. Properly used in this agricultural science context, "organic" refers to the methods grown and processed, not necessarily the chemical composition of the food.

In the industrial era, organic gardening reached a modest level of popularity in the United States in the 1950s. In the 1960s, environmentalists and the counterculture championed organic food, but it was only in the 1970s that a national marketplace for organic foods developed.

Early consumers interested in organic food would look for non-chemically treated, non-use of unapproved pesticides, fresh or minimally processed food. They mostly had to buy directly from growers. Later, "Know your farmer, know your food" became the motto of a new initiative instituted by the USDA in September 2009. Personal definitions of what constituted "organic" were developed through firsthand experience: by talking to farmers, seeing farm conditions, and farming activities. Small farms grew vegetables (and raised livestock) using organic farming practices, with or without certification, and the individual consumer monitored. Small specialty health food stores and co-operatives were instrumental to bringing organic food to a wider audience. As demand for organic foods continued to increase, high volume sales through mass outlets such as supermarkets rapidly replaced the direct farmer connection. Today, many large corporate farms have an organic division. However, for supermarket consumers, food production is not easily observable, and product labeling, like "certified organic," is relied upon. Government regulations and third-party inspectors are looked to for assurance.

In the 1970s, interest in organic food grew with the rise of the environmental movement, and was also spurred by food-related health scares like the concerns about Alar that arose in the mid-1980s.

Legal Definition

Organic food production is a self-regulated industry with government oversight in some countries, distinct from private gardening. Currently, the European Union, the United States, Canada, Japan, and many other countries require producers to obtain special certification based on government-defined standards in order to market food as organic within their borders. In the context of these regulations, foods marketed as organic are produced in a way that complies with organic standards set by national governments and international organic industry trade organizations.

In the United States, organic production is managed in accordance with the Organic Foods Production Act of 1990 (OFPA) and regulations in Title 7, Part 205 of the Code of Federal Regulations to respond to site-specific conditions by integrating cultural, biological, and mechanical practices that foster cycling of resources, promote ecological balance, and conserve biodiversity. If livestock are involved, the livestock must be reared with regular access to pasture and without the routine use of antibiotics or growth hormones.

The National Organic Program is in charge of the legal definition of organic in the United States.

Processed organic food usually contains only organic ingredients. If non-organic ingredients are present, at least a certain percentage of the food's total plant and animal ingredients must be organic (95% in the United States, Canada, and Australia). Foods claiming to be organic must be free of artificial food additives, and are often processed with fewer artificial methods, materials and conditions, such as chemical ripening, food irradiation, and genetically modified ingredients. Pesticides are allowed as long as they are not synthetic. However, under US federal organic standards, if pests and weeds are not controllable through management practices, nor via organic pesticides and herbicides, "a substance included on the National List of synthetic substances allowed for use in organic crop production may be applied to prevent, suppress, or control pests, weeds, or diseases." Several groups have called for organic standards to prohibit nanotechnology on the basis of the precautionary principle in light of unknown risks of nanotechnology. The use of nanotechnology-based products in the production of organic food is prohibited in some jurisdictions (Canada, the UK, and Australia) and is unregulated in others.

To be certified organic, products must be grown and manufactured in a manner that adheres to standards set by the country they are sold in:

- Australia: NASAA Organic Standard,

- Canada,

- European Union: EU-Eco-regulation.

  ○ Sweden: KRAV,

- ○ United Kingdom: DEFRA,

- ○ Poland: Association of Polish Ecology,

- ○ Norway: Debio Organic certification.

- India: National Program for Organic Production (NPOP).

- Indonesia: BIOCert, run by Agricultural Ministry of Indonesia.

- Japan: JAS Standards.

- Mexico: Consejo Nacional de Producción Orgánica, department of Sagarpa.

- New Zealand: there are three bodies; BioGro, AsureQuality, and OFNZ.

- United States: National Organic Program (NOP) Standards.

In the United States, there are four different levels or categories for organic labeling. 1) '100%' Organic: This means that all ingredients are produced organically. It also may have the USDA seal. 2) 'Organic': At least 95% or more of the ingredients are organic. 3) 'Made With Organic Ingredients': Contains at least 70% organic ingredients. 4) 'Less Than 70% Organic Ingredients': Three of the organic ingredients must be listed under the ingredient section of the label. In the U.S., the food label "natural" or "all natural" does not mean that the food was produced and processed organically.

Public Perception

There is widespread public belief that organic food is safer, more nutritious, and better tasting than conventional food,which has largely contributed to the development of an organic food culture. Consumers purchase organic foods for different reasons, including concerns about the effects of conventional farming practices on the environment, human health, and animal welfare.

The most important reason for purchasing organic foods seems to be beliefs about the products' health-giving properties and higher nutritional value. These beliefs are promoted by the organic food industry, and have fueled increased demand for organic food despite higher prices and difficulty in confirming these claimed benefits scientifically. Organic labels also stimulate the consumer to view the product as having more positive nutritional value.

Psychological effects such as the "halo" effect, which are related to the choice and consumption of organic food, are also important motivating factors in the purchase of organic food. The perception that organic food is low-calorie food or health food appears to be common.

In China the increasing demand for organic products of all kinds, and in particular milk, baby food and infant formula, has been "spurred by a series of food scares, the worst being the death of six children who had consumed baby formula laced with melamine" in 2009 and the 2008 Chinese milk scandal, making the Chinese market for organic milk the largest in the world as of 2014. A Pew Research Centre survey in 2012 indicated that 41% of Chinese consumers thought of food safety as a very big problem, up by three times from 12% in 2008.

Taste

There is no good evidence that organic food tastes better than its non-organic counterparts. There is evidence that some organic fruit is drier than conventionally grown fruit; a slightly drier fruit may also have a more intense flavor due to the higher concentration of flavoring substances.

Some foods, such as bananas, are picked when unripe, are cooled to prevent ripening while they are shipped to market, and then are induced to ripen quickly by exposing them to propylene or ethylene, chemicals produced by plants to induce their own ripening; as flavor and texture changes during ripening, this process may affect those qualities of the treated fruit.

Chemical Composition

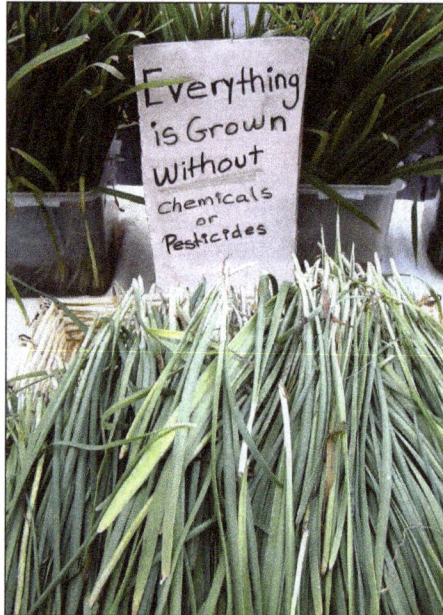

Organic vegetables at a farmers' market.

With respect to chemical differences in the composition of organically grown food compared with conventionally grown food, studies have examined differences in nutrients, antinutrients, and pesticide residues. These studies generally suffer from confounding variables, and are difficult to generalize due to differences in the tests that were done, the methods of testing, and because the vagaries of agriculture affect the chemical composition of food; these variables include variations in weather (season to season as well as place to place); crop treatments (fertilizer, pesticide, etc.); soil composition; the cultivar used, and in the case of meat and dairy products, the parallel variables in animal production. Treatment of the foodstuffs after initial gathering (whether milk is pasteurized or raw), the length of time between harvest and analysis, as well as conditions of transport and storage, also affect the chemical composition of a given item of food. Additionally, there is evidence that organic produce is drier than conventionally grown produce; a higher content in any chemical category may be explained by higher concentration rather than in absolute amounts.

Nutrients

Many people believe that organic foods have higher content of nutrients and thus are healthier than conventionally produced foods. However, scientists have not been equally convinced that this is the case as the research conducted in the field has not shown consistent results.

A 2009 systematic review found that organically produced foodstuffs are not richer in vitamins and minerals than conventionally produced foodstuffs. The results of the systematic review only showed a lower nitrogen and higher phosphorus content in organic produced compared to conventionally grown foodstuffs. Content of vitamin C, calcium, potassium, total soluble solids, copper, iron, nitrates, manganese, and sodium did not differ between the two categories.

A 2012 survey of the scientific literature did not find significant differences in the vitamin content of organic and conventional plant or animal products, and found that results varied from study to study. Produce studies reported on ascorbic acid (vitamin C), beta-carotene (a precursor for vitamin A), and alpha-tocopherol (a form of vitamin E) content; milk studies reported on beta-carotene and alpha-tocopherol levels. Few studies examined vitamin content in meats, but these found no difference in beta-carotene in beef, alpha-tocopherol in pork or beef, or vitamin A retinol in beef. The authors analyzed 11 other nutrients reported in studies of produce. A 2011 literature review found that organic foods had a higher micronutrient content overall than conventionally produced foods.

Similarly, organic chicken contained higher levels of omega-3 fatty acids than conventional chicken. The authors found no difference in the protein or fat content of organic and conventional raw milk.

A 2016 systematic review and meta-analysis found that organic meat had comparable or slightly lower levels of saturated fat and monounsaturated fat as conventional meat, but higher levels of both overall and n-3 polyunsaturated fatty acids. Another meta-analysis published the same year found no significant differences in levels of saturated and monounsaturated fat between organic and conventional milk, but significantly higher levels of overall and n-3 polyunsaturated fatty acids in organic milk than in conventional milk.

Anti-nutrients

The amount of nitrogen content in certain vegetables, especially green leafy vegetables and tubers, has been found to be lower when grown organically as compared to conventionally. When evaluating environmental toxins such as heavy metals, the USDA has noted that organically raised chicken may have lower arsenic levels. Early literature reviews found no significant evidence that levels of arsenic, cadmium or other heavy metals differed significantly between organic and conventional food products. However, a 2014 review found lower concentrations of cadmium, particularly in organically grown grains.

Phytochemicals

A 2014 meta-analysis of 343 studies on phytochemical composition found that organically grown crops had lower cadmiumand pesticide residues, and 17% higher concentrations of polyphenols than conventionally grown crops. Concentrations of phenolic acids, flavanones, stilbenes, flavones, flavonols, and anthocyanins were elevated, with flavanones being 69% higher. Studies on phytochemical composition of organic crops have numerous deficiencies, including absence of standardized measurements and poor reporting on measures of variability, duplicate or selective reporting of data, publication bias, lack of rigor in studies comparing pesticide residue levels in organic and conventional crops, the geographical origin of samples, and inconsistency of farming and post-harvest methods.

Pesticide Residues

The amount of pesticides that remain in or on food is called pesticide residue. In the United States, before a pesticide can be used on a food crop, the U.S. Environmental Protection Agency must determine whether that pesticide can be used without posing a risk to human health.

A 2012 meta-analysis determined that detectable pesticide residues were found in 7% of organic produce samples and 38% of conventional produce samples. This result was statistically heterogeneous, potentially because of the variable level of detection used among these studies. Only three studies reported the prevalence of contamination exceeding maximum allowed limits; all were from the European Union. A 2014 meta-analysis found that conventionally grown produce was four times more likely to have pesticide residue than organically grown crops.

The American Cancer Society has stated that no evidence exists that the small amount of pesticide residue found on conventional foods will increase the risk of cancer, although it recommends thoroughly washing fruits and vegetables. They have also stated that there is no research to show that organic food reduces cancer risk compared to foods grown with conventional farming methods.

The Environmental Protection Agency maintains strict guidelines on the regulation of pesticides by setting a tolerance on the amount of pesticide residue allowed to be in or on any particular food. Although some residue may remain at the time of harvest, residue tend to decline as the pesticide breaks down over time. In addition, as the commodities are washed and processed prior to sale, the residues often diminish further.

Bacterial Contamination

A 2012 meta-analysis determined that prevalence of *E. coli* contamination was not statistically significant (7% in organic produce and 6% in conventional produce). While bacterial contamination is common among both organic and conventional animal products, differences in the prevalence of bacterial contamination between organic and conventional animal products were also statistically insignificant.

Organic Meat Production Requirements

United States

Organic meat certification in the United States requires farm animals to be raised according to USDA organic regulations throughout their lives. These regulations require that livestock are fed certified organic food that contains no animal byproducts. Further, organic farm animals can receive no growth hormones or antibiotics, and they must be raised using techniques that protect native species and other natural resources. Irradiation and genetic engineering are not allowed with organic animal production. One of the major differences in organic animal husbandry protocol is the pasture rule: minimum requirements for time on pasture do vary somewhat by species and between the certifying agencies, but the common theme is to require as much time on pasture as possible and reasonable.

Health and Safety

There is little scientific evidence of benefit or harm to human health from a diet high in organic food, and conducting any sort of rigorous experiment on the subject is very difficult. A 2012 meta-analysis noted that "there have been no long-term studies of health outcomes of populations consuming predominantly organic versus conventionally produced food controlling for socioeconomic factors; such studies would be expensive to conduct." A 2009 meta-analysis noted that "most of the included articles did not study direct human health outcomes. In ten of the included studies (83%), a primary outcome was the change in antioxidant activity. Antioxidant status and activity are useful biomarkers but do not directly equate to a health outcome. Of the remaining two articles, one recorded proxy-reported measures of atopic manifestations as its primary health outcome, whereas the other article examined the fatty acid composition of breast milk and implied possible health benefits for infants from the consumption of different amounts of conjugated linoleic acids from breast milk." In addition, as discussed above, difficulties in accurately and meaningfully measuring chemical differences between organic and conventional food make it difficult to extrapolate health recommendations based solely on chemical analysis.

As of 2012, the scientific consensus is that while "consumers may choose to buy organic fruit, vegetables and meat because they believe them to be more nutritious than other food the balance of current scientific evidence does not support this view." The evidence of beneficial health effects of organic food consumption is scarce, which has led researchers to call for more long-term studies. In addition, studies that suggest that organic foods may be healthier than conventional foods face significant methodological challenges, such as the correlation between organic food consumption and factors known to promote a healthy lifestyle. When the American Academy of Pediatrics reviewed the literature on organic foods in 2012, they found that "current evidence does not support any meaningful nutritional benefits or deficits from eating organic compared with conventionally grown foods, and there are no well-powered human studies that directly demonstrate health benefits or disease protection as a result of consuming an organic diet."

Consumer Safety

Pesticide Exposure

The main difference between organic and conventional food products are the chemicals involved during production and processing. The residues of those chemicals in food products have dubious effects on human health. All food products on the market including those that contain residues of pesticides, antibiotics, growth hormones and other types of chemicals that are used during production and processing are said to be safe.

Claims of improved safety of organic food has largely focused on pesticide residues. These concerns are driven by the facts that: "(1) acute, massive exposure to pesticides can cause significant adverse health effects; (2) food products have occasionally been contaminated with pesticides, which can result in acute toxicity; and (3) most, if not all, commercially purchased food contains trace amounts of agricultural pesticides." However, as is frequently noted in the scientific literature: "What does not follow from this, however, is that chronic exposure to the trace amounts of pesticides found in food results in demonstrable toxicity. This possibility is practically impossible to study and quantify;" therefore firm conclusions about the relative safety of organic foods have been hampered by the difficulty in proper study design and relatively small number of studies directly comparing organic food to conventional food.

Additionally, the Carcinogenic Potency Project, which is a part of the US EPA's Distributed Structure-Searchable Toxicity (DSSTox) Database Network, has been systemically testing the carcinogenicity of chemicals, both natural and synthetic, and building a publicly available database of the results for the past ~30 years. Their work attempts to fill in the gaps in our scientific knowledge of the carcinogenicity of all chemicals, both natural and synthetic, as the scientists conducting the Project described:

> Toxicological examination of synthetic chemicals, without similar examination of chemicals that occur naturally, has resulted in an imbalance in both the data on and the perception of chemical carcinogens. Three points that we have discussed indicate that comparisons should be made with natural as well as synthetic chemicals.

1) The vast proportion of chemicals that humans are exposed to occur naturally. Nevertheless, the public tends to view chemicals as only synthetic and to think of synthetic chemicals as toxic despite the fact that every natural chemical is also toxic at some dose. The daily average exposure of Americans to burnt material in the diet is ~2000 mg, and exposure to natural pesticides (the chemicals that plants produce to defend themselves) is ~1500 mg. In comparison, the total daily exposure to all synthetic pesticide residues combined is ~0.09 mg. Thus, we estimate that 99.99% of the pesticides humans ingest are natural. Despite this enormously greater exposure to natural chemicals, 79% (378 out of 479) of the chemicals tested for carcinogenicity in both rats and mice are synthetic (that is, do not occur naturally).

2) It has often been wrongly assumed that humans have evolved defenses against the natural chemicals in our diet but not against the synthetic chemicals. However, defenses that animals have evolved are mostly general rather than specific for particular chemicals; moreover, defenses are generally inducible and therefore protect well from low doses of both synthetic and natural chemicals.

3) Because the toxicology of natural and synthetic chemicals is similar, one expects (and finds) a similar positivity rate for carcinogenicity among synthetic and natural chemicals. The positivity rate among chemicals tested in rats and mice is ~50%. Therefore, because humans are exposed to so many more natural than synthetic chemicals (by weight and by number), humans are exposed to an enormous background of rodent carcinogens, as defined by high-dose tests on rodents. We have shown that even though only a tiny proportion of natural pesticides in plant foods have been tested, the 29 that are rodent carcinogens among the 57 tested, occur in more than 50 common plant foods. It is probable that almost every fruit and vegetable in the supermarket contains natural pesticides that are rodent carcinogens.

While studies have shown via chemical analysis, as discussed above, that organically grown fruits and vegetables have significantly lower pesticide residue levels, the significance of this finding on actual health risk reduction is debatable as both conventional foods and organic foods generally have pesticide levels well below government established guidelines for what is considered safe. This view has been echoed by the U.S. Department of Agriculture and the UK Food Standards Agency.

A study published by the National Research Council in 1993 determined that for infants and children, the major source of exposure to pesticides is through diet. A study published in 2006 by Lu et al. measured the levels of organophosphorus pesticide exposure in 23 school children before and after replacing their diet with organic food. In this study it was found that levels of organophosphorus pesticide exposure dropped from negligible levels to undetectable levels when the children switched to an organic diet, the authors presented this reduction as a significant reduction in risk.

More specifically, claims related to pesticide residue of increased risk of infertility or lower sperm counts have not been supported by the evidence in the medical literature. Likewise the American Cancer Society (ACS) has stated their official position that "whether organic foods carry a lower risk of cancer because they are less likely to be contaminated by compounds that might cause cancer is largely unknown." Reviews have noted that the risks from microbiological sources or natural toxins are likely to be much more significant than short term or chronic risks from pesticide residues.

Microbiological Contamination

Organic farming has a preference for using manure as fertilizer, compared to conventional farming in general. This practise seems to imply an increased risk of microbiological contamination, such as *E. coli* O157:H7, from organic food consumption, but reviews have found little evidence that actual incidence of outbreaks can be positively linked to organic food production. The 2011 Germany E. coli O104:H4 outbreak, however, was blamed on organic farming of bean sprouts.

Environmental Safety

From an environmental perspective, fertilizing, overproduction and the use of pesticides in conventional farming has caused, and is causing, enormous damage worldwide to local ecosystems, biodiversity, groundwater and drinking water supplies, and sometimes farmer's health and fertility. Outcomes from organic farming, however, are uncertain for their environmental benefits, and have limits for transforming the food system, where reducing food waste and dietary changes might provide greater benefits.

Organic Beans

Organic beans are produced and processed without the use of synthetic fertilizers and pesticides. In 2008, over 2,600,000 acres (11,000 km²) of cropland were certified organic in the United States. Dry beans, snap beans, and soybeans were grown on 16,000 acres (65 km²), 5,200 acres (21 km²), and 98,000 acres (400 km²), respectively.

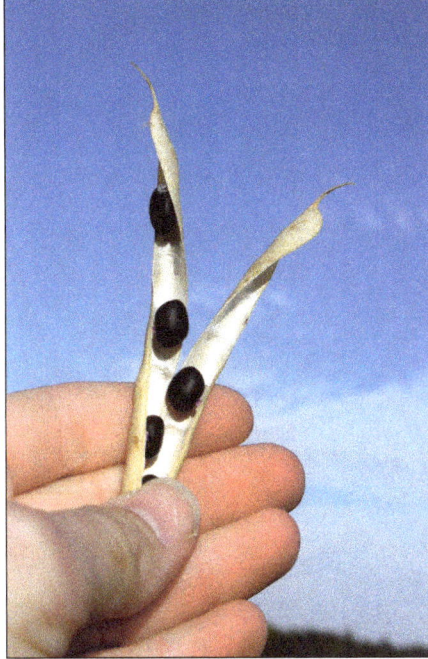
Black beans in the pod.

There are three major types of organic beans: dry beans, snap beans (also known as green beans), and soybeans. The mature seeds of dry beans (*Phaseolus vulgaris*) serve as a protein source in a variety of foods worldwide. Dry beans and snap beans are the same species, although dry beans are distinguished from snap beans (green beans) which are consumed as immature pods. Dry and snap beans also differ from soybeans (*Glycine max*), in which the seeds are consumed in a variety of processed forms such as tofu, soybean meal, oil, and fermented forms such as miso. Dry beans are divided into classes with a range of seed sizes, colors, and shapes. Examples include black beans, pintos, navy beans, small red beans, and kidney beans. Soybeans are usually divided into two groups, feed-grade and food-grade, with the food-grade including soybeans for processing, "tofu-beans", and edamame, the latter of which is eaten as immature beans.

Organic Certification

To sell beans as organic in the United States, producers must meet the requirements of the United States Department of Agriculture (USDA) National Organic Program (NOP) under the Organic Food Production Act of 1990. A third-party certifying agency, not part of the USDA, verifies that the producer has met the minimum requirements and may themselves have additional requirements. Prior to marketing crops as "certified organic", fields must be managed during a three-year transition phase using organic practices (that is, no synthetic pesticides or fertilizers). Detailed record keeping is essential for organic producers and these details are checked annually by the certifying agency who may perform a site inspection.

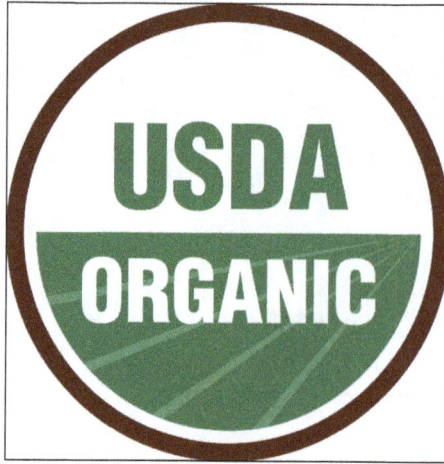

Labeling for products that meet the USDA-NOP standards.

Production Information

Organic Bean Producing Regions

Michigan is currently the top organic dry bean producing state in the U.S., accounting for 37% of the area and 47% of the sales. The majority of the area is located in the "Thumb" region of Michigan, with black beans being the most common class produced organically. Colorado and North Dakota are the second and third top producers, respectively, of organic dry beans in the United States.

Washington is the top producer of organic snap beans, followed by Michigan and California. In 2008, the total sales of organic snap beans totaled over 1.4 million dollars.

The value of organic soybeans in 2008 totaled more than $50 million in the U.S., with Minnesota, Iowa, and Michigan accounting for 46% of the total production.

Snap bean harvester.

Organic Bean Markets

Organically grown beans typically demand a price premium and as a consequence the marketing details differ from the conventional market. Organic bean producers often set up contracts with buyers prior to planting the crop. Buyers may also have special needs to consider when it comes to cleaning and processing beans in certified facilities, thereby making proximity to these certified facilities important.

Organic dry beans from the United States are marketed both nationally and abroad. In 2010, one of the markets for Michigan and Washington produced dry beans was the restaurant chain Chipotle Mexican Grill. Another market is Eden Foods Inc. Organic snap beans are marketed to a variety of companies, such as Gerber and Earth's Best, makers of organic baby food.

Organic soybeans have perhaps the most market potential of the three beans as they can be used as feed for organic animal production and processed into many different products.

Pest Management

Pest management in organic dry beans must be done without the use of synthetic herbicides, fungicides, insecticides, and other pesticides. For this reason, cultural, mechanical, and biological methods are the primary means of keeping pests under control.

Major Weed Pests

Rotary hoe.

Controlling weeds is the primary cost to agricultural production systems, including organic beans. Weeds can reduce the emergence, growth, and yield of beans as they compete for light, water, and nutrient resources. Weeds are grouped in many ways, such as by life cycle. In organic beans, the most problematic weeds are those with summer annual and perennial life cycles.

Summer Annuals

Summer annuals are weeds that germinate in the spring or summer, produce seed in the late summer to fall, and then die. Their life cycle is complete in one year.

- Common lambsquarters: Broadleaf weed that emerges over an extended period of time, grows rapidly, produces large quantities of seed that can remain dormant in the soil for decades.

- Common ragweed: Broadleaf weed that emerges over an extended period of time, grows rapidly, and is tolerant to control by heat treatments (like a propane flamer).

- Pigweed (*Amaranthus*): Broadleaf weed that grows rapidly and is a prolific producer of seed.

- Foxtail (*Setaria*) species: Grass weed that grows in clumps and emerges after common lambsquarters and common ragweed.

- Crabgrass (*Digitaria*) species: Grass weed with a prostrate growth habit.

Perennials

Perennial weeds can live for multiple growing seasons, usually due to hardy root stocks.

- Canada thistle: Broadleaf weed with roots that can grow to a depth 15 feet (4.6 m) in the soil profile. Tillage often promotes the spread of Canada thistle by dispersing segments of the root.

- Perennial sowthistle: Broadleaf weed with yellow flowers similar to dandelion that can reproduce by seed or rhizome.

- Quackgrass: Grass weed that reproduces primarily via rhizome. Seeds are short lived in the soil (2–4 years).

Example of weed management: Several types of tractor-pulled implements such as a rotary hoe and/or one of various types of cultivator/cultivators can be used to mechanically remove weeds from organic fields.

Major Insect Pests

Insects are another contributor to yield loss in organic beans. There are several examples of insects that affect organic bean systems including:

- Bean and potato leafhoppers- sucking insect that causes bean leaves to yellow at the tips and on the edges and can cause stunting.

- Western bean cutworm: Larvae of this insect feed on the pods of beans reducing yield and quality.

- Soybean aphid: Sap-sucking insect that can form large colonies on soybean. Soybean aphids can also transmit viruses from plant to plant while feeding.

- Seedcorn maggot: Feed on dry and snap bean seed and seedlings. The risk of feeding is increased when temperatures are cool and the soil is wet.

Western bean cutworm damage to dry bean.

Example of insect management: *Bacillus thuringiensis* (Bt) is a bacterium that when ingested by susceptible insects is lethal. This product is referred to as a microbial insecticide of which there are four subspecies and over 100 commercial products available for use in organic systems.

Major Pathogens

Pathogens can take the form of fungi, bacteria, viruses, and nematodes. Each can act as a disruptor to plant growth and development and can have negative effects on organic bean yields. Examples of bean pathogens include:

- White mold: A fungus with a wide host range. A symptom of later stages of infection is white cottony hyphal growth on bean stems, leaves, and pods.

- Anthracnose: Refers to disease caused by many different fungi. Anthracnose spores can be seedborne or overwinter on leaf litter.

- Soybean cyst nematode (soybean specific): Can cause up to 50% yield loss in soybeans due to root feeding.

- Bean common and soybean mosaic viruses: Virus that occurs on bean seed and can have negative impacts on yield and bean quality. Both viruses can be transmitted from plant to plant by aphids.

Example of disease management: The onset of certain diseases, such as white mold, can be reduced by planting in rows wide enough to allow adequate air movement which can reduce humidity and decrease drying time after precipitation events. If plant material is allowed to dry quickly it will not have the 9 to 48 hours of continuous moisture on the leaf surface required for the white mold infection to occur.

Organically Approved Pesticides

Some naturally derived chemical products are permitted for use under organic production. All materials have to be reviewed and approved by the Organic Materials Review Institute (OMRI). Examples of organically allowable pesticides that can be used to produce organic beans include:

- Neem oil: Used as an insecticide.

- Acetic acid: Used as a herbicide.

- Pyrethrum: Broad spectrum insecticide derived from Chrysanthemum.

- Bacillus subtilis: Used as a fungicide.

- Bacillus thuringiensis: Used as an insecticide.

Fertility

Because synthetic fertilizers are not permitted in organic production systems, fertility must come from sources such as:

- Green manures (cover crops).

- Composts and compost teas.

- Livestock manures.

- Rock minerals- such as limestone, gypsum, and rock phosphate.

References

- United States Department of Agriculture - Economic Research Service. "Table 2. U.S. certified organic farmland acreage, livestock numbers, and farm operations". ERS/USDA Data- Organic Production. Retrieved 2 March2011

- Organic-farming, topic: britannica.com, Retrieved 5 January, 2019

- Winter, Carl K.; Davis, Sarah F. (November 2006). "Organic Foods". Journal of Food Science. 71 (9): R117–R124. Doi:10.1111/j.1750-3841.2006.00196.x

- Organic-farming-benefits: conserve-energy-future.com, Retrieved 6 February, 2019

- "Safeguarding the Environment: Canadian Aquaculture Industry Alliance". Www.aquaculture.ca. Archived from the original on 2016-03-23. Retrieved 2016-04-22

- Organic-farming-in-india: techgape.com, Retrieved 7 March, 2019

- Wickramasinghe, L.P., Harris, S., Jones, G., Vaughan, N., 2003. Bat activity and species richness on organic and conventional farms: impact of agricultural intensification. Journal of Applied Ecology 40, 984–993

- Orgfarm-principles, org-farm: agritech.tnau.ac.in, Retrieved 8 April, 2019

- Biello, David (25 April 2012). "Will Organic Food Fail to Feed the World?". Scientific American. Scientific American. Retrieved 8 May 2015

- Pelletier, N; Tyedmers, P (2007), "Feeding farmed salmon: Is organic better?", Aquaculture, 272 (2): 399–416, doi:10.1016/j.aquaculture.2007.06.024

- John Jeavons, How to Grow More Vegetables: And Fruits, Nuts, Berries, Grains, and Other Crops Than You Ever Thought Possible on Less Land Than You Can Imagine ISBN 1-58008-233-5; "Biointensive agriculture"

- Blair, Robert. (2012). Organic Production and Food Quality: A Down to Earth Analysis. Wiley-Blackwell, Oxford, UK. Pages 72, 223. ISBN 978-0-8138-1217-5

- Caldwell, B, E. B. Rosen, E. Sideman, A. Shelton, and C. Smart (2005). Resource guide for organic insect and disease management. Geneva, NY: New York Agricultural Experiment Station. P. 169. ISBN 0-9676507-2-0

2
Methods and Practices in Organic Farming

The most common methods and practices of organic farming are crop-rotation, veg-an-organic gardening and composting. These measures make use of the natural environment to enhance productivity. The topics elaborated in this chapter will help in gaining a better understanding of these practices and methods of organic farming.

Crop Rotation

Crop rotation essentially involves moving plants of the same kind to new ground every year When you grow a crop in the same ground year after year, specific pests readily increase in numbers. And when some pests, such as potato eelworm, are allowed to proliferate they can remain a problem for years to come and prevent further potato growing. That would be a gardening catastrophy indeed.

But you can prevent these pests becoming a problem by moving your crop to another site in your garden. So when the pest eggs hatch the food plant needed to sustain them has gone. For example, with 4 different crops on 4 plots, each plot grows the same crop only once every 4 years. That's really simple and, neglecting rotation is just too risky.

And there's another advantage to crop rotation. Because, crops of a similar kind with similar root systems take the same nutrients from the same volume of soil. Whereas different crop types can take different nutrients at different soil depths. Some crops can even add nutrients to the soil to help make up for nutrient loss.

Therefore rotating your crops makes maximum use of soil nutrients before they need replacing and even allows some nutrient replenishment. However, it is important to remember that continuous end-to-end growing will cause a loss in productivity. Therefore your organic crop rotation should also allow for rest periods, including growing green manures and the application of composted animal manure.

Guidelines for Organic Crop Rotation

- Grow plants of the same family together and follow them with plants of a different plant family of course.

- But, avoid large monocultures which can also attract pests. Interplant rows with companion crops such as spring onion in between carrots. This is effective against carrot root fly.

- After crops that are preceded by gross feeding, use "nitrogen lifting" green manures to mop surplus nutrients if there's no crop to follow on.

- Use beans or a "nitrogen fixing" green manure before and after brassicas.

- Use one of: Animal manure, garden compost and green manure every year and roughly rotate which of the three is applied to a particular bed.

- After digging in green manures follow them with large seeds such as beans, or by direct planting out e.g. seed potatoes, onion sets, plants grown in modules etc.

- To maintain fertility during continuous production it is important to supplement nutrients with organic fertilizers.

- A long rotation cycle - 6 years or so - is important in protecting crops like potatoes. Growing in tubs helps to give the small gardens a rest.

- Note also that dried bagged manure can be used immediately like an organic fertilizer.

Vegan-organic Gardening

Vegan organic gardening and farming is the organic cultivation and production of food crops and other crops with a minimal amount of exploitation or harm to any animal Vegan-organic gardening avoids not only the use of toxic sprays and chemicals, but also manures and animal remains. Just as vegans avoid animal products in the rest of our lives, we also avoid using animal products in the garden, as fertilizers such as blood and bone meal, slaughterhouse sludge, fish emulsion, and manures are sourced from industries that exploit and enslave sentient beings. As these products may carry dangerous diseases that breed in intensive animal production operations, vegan-organic gardening is also a safer, healthier way to grow our food.

In veganic growing situations, soil fertility is maintained using vegetable compost, green manures, crop rotation, mulching, and other sustainable, ecological methods.

Occasional use of lime, gypsum, rock phosphorus, dolomite, rock dusts and rock potash can be helpful, but we try not to depend on these fertilizers as they are non-renewable resources.

Soil conditioners and fertilizers that are vegan-organic and ecologically sustainable include hay mulch, wood ash, composted organic matter (fruit/vegetable peels, leaves and grass clippings), green manures/nitrogen-fixing cover crops (fava beans/clover/alfalfa/lupines), liquid feeds (such as comfrey or nettles), and seaweed (fresh, liquid or meal) for trace elements.

A border of marigolds helps to deter certain insects, and they also have a root system that improves the soil.

Composted Organic Matter

A compost pile consists of food waste such as fruit and vegetable rinds, that is covered by course material like leaves or grass clippings. The object is to create layers of food material alternating with covering material to allow aeration. When a bin is full, the pile is flipped and covered by black plastic or weed mat to protect it from rain and create heat. It can be flipped again after a period of time, so the bottom becomes the top. Cover again and within a couple of months, depending on the climate, nature's master recycling plan will have taken its course and you will have vitamin-rich soil.

Green Manures

Green Manure is a cover crop of plants, which is grown with the specific purpose of being tilled into the soil. Fast-growing plants such as wheat, oats, rye, vetch, or clover, can be grown as cover crops between gardening seasons then tilled into the garden as it is prepared for the next planting. Green manure crops absorb and use nutrients from the soil that might otherwise be lost through leaching, then return these nutrients to the soil when they are tilled under. The root system of cover crops improves soil structure and helps prevent erosion. Nitrogen-fixing crops such as vetch, peas, broad beans (fava beans) and crimson clover add nitrogen to the soil as they are turned under and decompose. Cover crops also help reduce weed growth during the fall and winter months.

Liquid Feeds such as Comfrey or Nettles

Fill a container with grass cuttings, nettles, weed or comfrey leaves. Cover with water at a rate of one part brew to three parts water. Cover the container, and leave for two to four weeks. Preferably strain out (through an old stocking) the weed seeds and plant material that will block up the spout of your watering-can. Nettles give the best multi-purpose feed and comfrey alone will give a feed rich in potash.

Hay Mulches

Using a thick layer of hay to cover the earth feeds the soil with organic matter as it breaks down. It also suppresses weeds and encourages worms to live in your soil. When putting gardens to sleep over the winter, cover them with a very thick layer of hay mulch.

Seaweed

Used for trace elements. Seaweed is best harvested fresh from the sea as opposed to washed up and sitting on beaches. Some veganic gardeners use bulk spirulina or kelp meal (used for potash and trace minerals).

Worm Castings: Vermiculture, Vermicastings and Vermicomposting

Re-establish natural worm populations in your garden. Composting worms love cool, damp and dark environments (like under black weed mat or a thick layer of hay mulch), and will breed optimally when these conditions are maintained. Worm castings are a rich, all-natural source of organic matter with lots of nutrients and moisture-holding capabilities. Earthworm castings are known to have an extraordinary effect on plant life. They improve the soil structure and increase fertility.

Lime

The primary purpose for using lime in the garden is to reduce the acidity of the soil, otherwise known as raising the pH level or 'sweetening the soil'. Most plants prefer a fairly neutral soil for optimum growth. You can have your soil tested to see if it is acidic or alkaline. Lime also enriches the soil with calcium and magnesium. Calcium is essential for strong plant growth and aids in the absorption of other nutrients. Lime can also be used for breaking up heavy clay soil.

Gypsum

Gypsum is also used where more calcium is needed, but unlike Lime, it enriches the soil without raising the pH level.

Neem

Known as the wonder tree in India, Neem has been in use for centuries in Indian agriculture as the best natural pest repellent and organic fertilizer with insect sterilization properties.

EM Bokashi

Bokashi is a Japanese term that means 'fermented organic matter'. EM means Effective Micro-organisms and consists of mixed cultures of naturally occurring, beneficial micro-organisms such as lactic acid bacteria, yeast, photosynthetic bacteria and actinomycetes. It is a bran-based material that has been fermented with EM liquid concentrate and dried for storage. Add to compost to aid in the fermentation of the organic matter. EM Bokashi should be stored in a warm, dry place out of direct sunlight.

Green Sand

A soil amendment and fertilizer. It is mined from deposits of minerals that were originally part of the ocean floor. It is a natural source of potash, as well as iron, magnesium, silica and as many as 30 other trace minerals. It may also be used to loosen heavy clay soils. It has the consistency of sand but has 10 times the ability to absorb moisture.

Alfalfa Meal, Flax Seed Meal, Cottonseed Meal and Soya Meal

Sources of nitrogen.

Epsom Salts

An excellent source of magnesium.

Dolomite

A finely ground rock dust which is the preferred source of calcium and magnesium.

Rock Phosphate

Phosphorus is an essential element for plant and animal nutrition. It is mined in the form of phosphate rock, which formed in oceans in the form of calcium phosphate called phosphorite. The primary mineral in phosphate rock is apatite.

Rock Dusts

Used to re-mineralize soil that has become depleted through industrial and agricultural practices. It releases slowly into the soil and can be applied directly, in combination with other fertilizers, or added to the compost. These products have a highly stimulating effect on microbial activity.

Rock Potash

Potassium is an essential nutrient that enhances flower and fruit production and helps 'harden' foliage to make it less susceptible to disease. Rock potash is very slow-acting. It releases gradually as it weathers, which can take years. Use it when preparing soil before planting.

Composting

Composting is a form of waste disposal where organic waste decomposes naturally under oxygen-rich conditions. Although all waste will eventually decompose, only certain waste items are considered compostable and should be added to compost containers. Compost organic matter that has been decomposedin a process called composting. This process recycles various organic materials otherwise regarded as waste products and produces a soil conditioner.

Compost is rich in nutrients. It is used, for example, in gardens, landscaping, horticulture, urban agriculture and organic farming. The compost itself is beneficial for the land in many ways, including as a soil conditioner, a fertilizer, addition of vital humus or humic acids, and as a natural pesticide for soil. In ecosystems, compost is useful for erosion control, land and stream reclamation, wetland construction, and as landfill cover.

Community-level composting in a rural area in Germany.

At the simplest level, the process of composting requires making a heap of wet organic matter (also called green waste), such as leaves, grass, and food scraps, and waiting for the materials to break down into humus after a period of months. However, composting also can take place as a multi-step, closely monitored process with measured inputs of water, air, and carbon- and nitrogen-rich materials. The decomposition process is aided by shredding the plant matter, adding water and ensuring proper aeration by regularly turning the mixture when open piles or "windrows" are used. Earthworms and fungi further break up the material. Bacteria requiring oxygen to function (aerobic bacteria) and fungi manage the chemical process by converting the inputs into heat, carbon dioxide, and ammonium.

Fundamentals

Composting is an aerobic method (meaning that it requires the presence of air) of decomposing organic solid wastes. It can therefore be used to recycle organic material. The process involves decomposition of organic material into a humus-like material, known as compost, which is a good fertilizer for plants. Composting requires the following three components: human management, aerobic conditions, development of internal biological heat.

Home compost barrel.

Materials in a compost pile.

Food scraps compost heap.

Composting organisms require four equally important ingredients to work effectively:

- Carbon: For energy; the microbial oxidation of carbon produces the heat, if included at suggested levels. High carbon materials tend to be brown and dry.

- Nitrogen: To grow and reproduce more organisms to oxidize the carbon. High nitrogen materials tend to be green (or colorful, such as fruits and vegetables) and wet.

- Oxygen: For oxidizing the carbon, the decomposition process.

- Water: In the right amounts to maintain activity without causing anaerobic conditions.

Certain ratios of these materials will provide microorganisms to work at a rate that will heat up the pile. Active management of the pile (e.g. turning) is needed to maintain sufficient supply of oxygen and the right moisture level. The air/water balance is critical to maintaining high temperatures (135°-160 °F/50°-70 °C) until the materials are broken down.

The most efficient composting occurs with an optimal carbon:nitrogen ratio of about 25:1. Hot container composting focuses on retaining the heat to increase decomposition rate and produce compost more quickly. Rapid composting is favored by having a C/N ratio of ~30 or less. Above 30 the substrate is nitrogen starved, below 15 it is likely to outgas a portion of nitrogen as ammonia.

Nearly all plant and animal materials have both carbon and nitrogen, but amounts vary widely, with characteristics noted above (dry/wet, brown/green). Fresh grass clippings have an average ratio of about 15:1 and dry autumn leaves about 50:1 depending on species. Mixing equal parts by volume approximates the ideal C:N range. Few individual situations will provide the ideal mix of materials at any point. Observation of amounts, and consideration of different materials as a pile is built over time, can quickly achieve a workable technique for the individual situation.

Microorganisms

With the proper mixture of water, oxygen, carbon, and nitrogen, micro-organisms are able to break down organic matter to produce compost. The composting process is dependent on micro-organisms to break down organic matter into compost. There are many types of microorganisms found in active compost of which the most common are:

- Bacteria: The most numerous of all the microorganisms found in compost. Depending on the phase of composting, mesophilic or thermophilic bacteria may predominate.

- Actinobacteria: Necessary for breaking down paper products such as newspaper, bark, etc.

- Fungi: Molds and yeast help break down materials that bacteria cannot, especially lignin in woody material.

- Protozoa: Help consume bacteria, fungi and micro organic particulates.

- Rotifers: Rotifers help control populations of bacteria and small protozoans.

In addition, earthworms not only ingest partly composted material, but also continually re-create aeration and drainage tunnels as they move through the compost.

Phases of Composting

Under ideal conditions, composting proceeds through three major phases:

- An initial, mesophilic phase, in which the decomposition is carried out under moderate temperatures by mesophilic microorganisms.

- As the temperature rises, a second, thermophilic phase starts, in which the decomposition is carried out by various thermophilic bacteria under high temperatures.

- As the supply of high-energy compounds dwindles, the temperature starts to decrease, and the mesophiles once again predominate in the maturation phase.

Three years old household compost.

Slow and Rapid Composting

There are many proponents of rapid composting that attempt to correct some of the perceived problems associated with traditional, slow composting. Many advocate that compost can be made in 2 to 3 weeks. Many such short processes involve a few changes to traditional methods, including smaller, more homogenized pieces in the compost, controlling carbon-to-nitrogen ratio (C:N) at 30 to 1 or less, and monitoring the moisture level more carefully. However, none of these parameters differ significantly from the early writings of compost researchers, suggesting that in fact modern composting has not made significant advances over the traditional methods that take a few months to work. For this reason and others, many scientists who deal with carbon transformations are sceptical that there is a "super-charged" way to get nature to make compost rapidly.

Both sides may be right to some extent. The bacterial activity in rapid high heat methods breaks down the material to the extent that pathogens and seeds are destroyed, and the original feedstock is unrecognizable. At this stage, the compost can be used to prepare fields or other planting areas. However, most professionals recommend that the compost be given time to cure before using in a nursery for starting seeds or growing young plants. The curing time allows fungi to continue the decomposition process and eliminating phytotoxic substances.

An alternative approach is anaerobic fermentation, known as bokashi. It retains carbon bonds, is faster than decomposition, and for application to soil requires only rapid but thorough aeration rather than curing. It depends on sufficient carbohydrates in the treated material.

Pathogen Removal

Composting can destroy pathogens or unwanted seeds. Unwanted living plants (or weeds) can be discouraged by covering with mulch/compost. The "microbial pesticides" in compost may include thermophiles and mesophiles.

Thermophilic (high-temperature) composting is well known to destroy many seeds and nearly all types of pathogens (exceptions may include prions). The sanitizing qualities of (thermophilic) composting are desirable where there is a high likelihood of pathogens, such as with manure.

Materials that can be Composted

Composting is a process used for resource recovery. It can recycle an unwanted by-product from another process (a waste) into a useful new product.

Organic Solid Waste

A large compost pile that is steaming with the heat generated by thermophilic microorganisms.

Composting is a process for converting decomposable organic materials into useful stable products. Therefore, valuable landfill space can be used for other wastes by composting these materials rather than dumping them on landfills. It may however be difficult to control inert and plastics contamination from municipal solid waste.

Co-composting is a technique that processes organic solid waste together with other input materials such as dewatered fecal sludge or sewage sludge.

Industrial composting systems are being installed to treat organic solid waste and recycle it rather than landfilling it. It is one example of an advanced waste processing system. Mechanical sorting of mixed waste streams combined with anaerobic digestion or in-vessel composting is called mechanical biological treatment. It is increasingly being used in developed countries due to regulations controlling the amount of organic matter allowed in landfills. Treating biodegradable waste before it enters a landfill reduces global warming from fugitive methane; untreated waste breaks down anaerobically in a landfill, producing landfill gas that contains methane, a potent greenhouse gas.

Animal Manure and Bedding

On many farms, the basic composting ingredients are animal manure generated on the farm and bedding. Straw and sawdust are common bedding materials. Non-traditional bedding materials are also used, including newspaper and chopped cardboard. The amount of manure composted on a livestock farm is often determined by cleaning schedules, land availability, and weather conditions. Each type of manure has its own physical, chemical, and biological characteristics. Cattle and horse manures, when mixed with bedding, possess good qualities for composting. Swine manure, which is very wet and usually not mixed with bedding material, must be mixed with straw or similar raw materials. Poultry manure also must be blended with carbonaceous materials - those low in nitrogen preferred, such as sawdust or straw.

Human Excreta and Sewage Sludge

Human excreta can also be added as an input to the composting process since human excreta is a nitrogen-rich organic material. It can be either composted directly, like in composting toilets, or indirectly (as sewage sludge), after it has undergone treatment in a sewage treatment plant.

Urine can be put on compost piles or directly used as fertilizer. Adding urine to compost can increase temperatures and therefore, increase its ability to destroy pathogens and unwanted seeds. Unlike feces, urine does not attract disease-spreading flies (such as houseflies or blowflies), and it does not contain the most hardy of pathogens, such as parasitic wormeggs. Urine usually does not smell for long, particularly when it is fresh, diluted, or put on sorbents.

Uses

Compost can be used as an additive to soil, or other matrices such as coir and peat, as a tilth improver, supplying humus and nutrients. It provides a rich *growing medium*, or a porous, absorbent material that holds moisture and soluble minerals, providing the support and nutrients in which plants can flourish, although it is rarely used alone, being primarily mixed with soil, sand, grit, bark chips, vermiculite, perlite, or clay granules to produce loam. Compost can be tilled directly into the soil or growing medium to boost the level of organic matter and the overall fertility of the soil. Compost that is ready to be used as an additive is dark brown or even black with an earthy smell.

Generally, direct seeding into a compost is not recommended due to the speed with which it may dry and the possible presence of phytotoxins in immature compost that may inhibit germination, and the possible tie up of nitrogen by incompletely decomposed lignin. It is very common to see blends of 20–30% compost used for transplanting seedlings at cotyledon stage or later.

Compost can be used to increase plant immunity to diseases and pests.

Composting Technologies

Various approaches have been developed to handle different ingredients, locations, throughput and applications for the composted product.

Industrial-scale

Industrial-scale composting can be carried out in the form of in-vessel composting, aerated static pile composting, vermicomposting, or windrow composting.

Vermicomposting

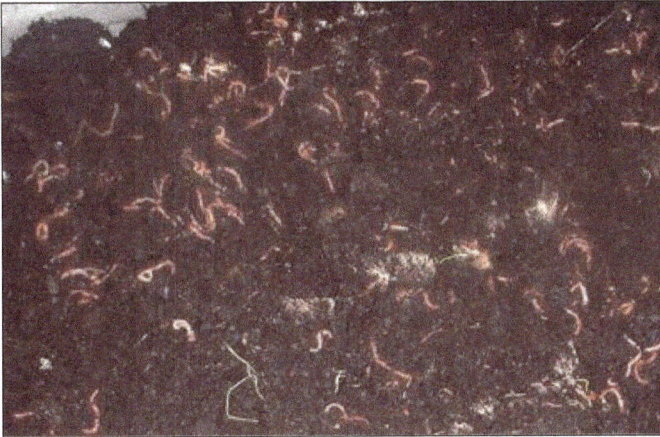

Worms in a bin being harvested.

Vermicompost is the product or process of organic material degradation using various species of worms, usually red wigglers, white worms, and earthworms, to create a heterogeneous mixture of decomposing vegetable or food waste (excluding meat, dairy, fats, or oils), bedding materials, and vermicast. Vermicast, also known as worm castings, worm humus or worm manure, is the end-product of the breakdown of organic matter by species of earthworm. Vermicomposting can also be applied for treatment of sewage sludge.

Composting Toilets

A composting toilet collects human excreta. These are added to a compost heap that can be located in a chamber below the toilet seat. Sawdust and straw or other carbon rich materials are usually added as well. Some composting toilets do not require water or electricity; others may. If they do not use water for flushing they fall into the category of dry toilets. Some composting toilet designs use urine diversion, others do not. When properly managed, they do not smell. The composting process in these toilets destroys pathogens to some extent. The amount of pathogen destruction depends on the temperature (mesophilic or thermophilic conditions) and composting time.

Composting toilet with a seal in the lid in Germany.

Composting toilets with a large composting container (of the type Clivus Multrum and derivations of it) are popular in United States, Canada, Australia, New Zealand and Sweden. They are available as commercial products, as designs for self builders or as "design derivatives" which are marketed under various names.

Black Soldier Fly Larvae

Black soldier fly (*Hermetia illucens*) larvae are able to rapidly consume large amounts of organic material when kept at around 30 °C. Black soldier fly larvae can reduce the dry matter of the organic waste by 73% and convert 16-22% of the dry matter in the waste to biomass. The resulting compost still contains nutrients and can be used for biogasproduction, or further traditional composting or vermicomposting. The larvae are rich in fat and protein, and can be used as, for example, animal feed or bio-diesel production. Enthusiasts have experimented with a large number of different waste products.

Bokashi

Bokashi is not composting as defined earlier, rather an alternative technology. It ferments (rather than decomposes) the input organic matter and feeds the result to the soil food web (rather than producing a soil conditioner). The process involves adding *Lactobacilli* to the input in an airtight container kept at normal room temperature. These bacteria ferment carbohydrates to lactic acid, which preserves the input. After this is complete the preserve is mixed into soil, converting the lactic acid to pyruvate, which enables soil life to consume the result.

Bokashi is typically applied to food waste from households, workplaces and catering establishments, because such waste normally holds a good proportion of carbohydrates; it is also applied to other organic waste by supplementing carbohydrates. Household containers ("bokashi bins") typically give a batch size of 5-10 kilograms, accumulated over a few weeks. In horticultural settings batches can be orders of magnitude greater.

Inside a recently started bokashi bin. Food scraps are raised on a perforated plate (to drain runoff) and are partly covered by a layer of bran inoculated with *Lactobacilli*.

Bokashi offers several advantages:

- Fermentation retains all the original carbon and energy. (In comparison, composting loses at least 50% of these and 75% or more in amateur use; composting also loses nitrogen, a macronutrient of plants, by emitting ammonia and the potent greenhouse gas nitrous oxide).

- Virtually the full range of food waste is accepted, without the exclusions of composting. The exception is large bones.

- Being airtight, the container inherently traps smells, and when opened the smell of fermentation is far less offensive than decomposition. Hence bokashi bins usually operate indoors, in or near kitchens.

- Similarly, the container neither attracts insect pests nor allows them ingress.

- The process is inherently hygienic because lactic acid is a natural bactericide- and anti-pathogen; even its own fermentation is self-limiting.

- Both preservation and consumption complete within a few weeks rather than months.

- The preserve can be stored until needed, for example if ground is frozen or waterlogged.

- The increased activity of the soil food web improves the soil texture, especially by worm action - in effect this is in-soil vermicomposting.

The importance of the first advantage should not be underestimated: the mass of any ecosystem depends on the energy it captures. Plants depend upon the soil ecosystem making nutrients available within soil water. Therefore, the richer the ecosystem, the richer the plants. (Plants can also take up nutrients from added chemicals, but these are at odds with the purpose of composting).

Other Systems at Household Level

Hügelkultur

An almost completed Hügelkultur bed; the bed
does not have soil on it yet.

The practice of making raised garden beds or mounds filled with rotting wood is also called *hügelkultur* in German. It is in effect creating a nurse log that is covered with soil.

Benefits of *hügelkultur* garden beds include water retention and warming of soil. Buried wood acts like a sponge as it decomposes, able to capture water and store it for later use by crops planted on top of the *hügelkultur* bed.

Compost Tea

Compost teas are defined as water extracts leached from composted materials. Compost teas are generally produced from adding one volume of compost to 4-10 volumes of water, but there has also been debate about the benefits of aerating the mixture. Field studies have shown the benefits of adding compost teas to crops due to the adding of organic matter, increased nutrient availability and increased microbial activity. They have also been shown to have an effect on plant pathogens.

Worm Hotels

Worm Hotels accommodate useful worm in ideal conditions

Worm Hotel in Amsterdam.

Related Technologies

Organic ingredients intended for composting can also be used to generate biogas-through anaerobic digestion. This process stabilizes organic material. The residual material, sometimes in combination with sewage sludge can be treated by a composting process before selling or giving away the compost.

Regulations

There are process and product guidelines in Europe that date to the early 1980s (Germany, the Netherlands, Switzerland) and only more recently in the UK and the US. In both these countries, private trade associations within the industry have established loose standards, some say as a stop-gap measure to discourage independent government agencies from establishing tougher consumer-friendly standards.

The USA is the only Western country that does not distinguish sludge-source compost from green-composts, and by default in the USA 50% of states expect composts to comply in some manner with the federal EPA 503 rule promulgated in 1984 for sludge products. Compost is regulated in Canada and Australia as well.

Many countries such as Wales and some individual cities such as Seattle and San Francisco require food and yard waste to be sorted for composting (San Francisco Mandatory Recycling and Composting Ordinance).

Edmonton Composting Facility.

Large-scale composting systems are used by many urban areas around the world.

- The world's largest municipal co-composter for municipal solid waste (MSW) is the Edmonton Composting Facility in Edmonton, Alberta, Canada, which turns 220,000 tonnes of municipal solid waste and 22,500 dry tonnes of sewage sludge per year into 80,000 tonnes of compost. The facility is 38,690 m² (416,500 sq.ft.) in area, equivalent to 4½ Canadian football fields, and the operating structure is the largest stainless steel building in North America.

- In 2006, Qatar awarded Keppel Seghers Singapore, a subsidiary of Keppel Corporation, a contract to begin construction on a 275,000 tonne/year anaerobic digestion and composting plant licensed by Kompogas Switzerland. This plant, with 15 independent anaerobic digesters, will be the world's largest composting facility once fully operational in early 2011 and forms part of Qatar's Domestic Solid Waste Management Centre, the largest integrated waste management complex in the Middle East.

- Another large municipal solid waste composter is the Lahore Composting Facility in Lahore, Pakistan, which has a capacity to convert 1,000 tonnes of municipal solid waste per day into compost. It also has a capacity to convert substantial portion of the intake into refuse-derived fuel (RDF) materials for further combustion use in several energy consuming industries across Pakistan, for example in cement manufacturing companies where it is used to heat cement kilns. This project has also been approved by the Executive Board of the United Nations Framework Convention on Climate Change for reducing methane emissions, and has been registered with a capacity of reducing 108,686 tonnes carbon dioxide equivalent per annum.

- Kew Gardens in London has one of the biggest non-commercial compost heaps in Europe.

- Compost is used as a soil amendment in organic farming.

3
Diverse Types of Organic Fertilizers

The fertilizers derived from animal matter, vegetable matter, animal excreta and human excreta are known as organic fertilizers. They can be used as a substitute of chemical fertilizers. This chapter has been carefully written to provide an easy understanding of the varied facets of organic fertilizers as well their advantages and disadvantages.

An organic fertilizer is a plant fertilizer that is derived from organic sources. Organic fertilizers can range from organic compost to cow manure, but they must be derived from all-organic sources.

Chicken droppings from an organic farm would be considered an organic fertilizer. Additional examples of organic fertilizer sources include kelp, guano, bone and blood meals, molasses, and fish emulsions.

Cow manure from a conventional farm that uses commercial fertilizers and pesticides would not be considered organic.

Organic fertilizers differ from chemical fertilizers in that they feed your plants while building a healthy soil. They are considered the more environmentally friendly option. Soils with plenty of organic material remain loose and light, retain more moisture and nutrients, and foster growth of soil micro-organisms that promote healthier plants and root development.

If a lot of synthetic chemicals are added to a soil, it eventually loses its organic matter percentage and micro-biotic activity, which leads to an unhealthy soil. As organic material is used up, the soil structure deteriorates; it becomes harder and less able to hold water and nutrients.

For organic gardeners, creating a living soil rich in humus and nutrients is the key to growing great fruits and vegetables, abundant flowers, and long-lived ornamental trees and shrubs. The overall fertility and viability of the soil, rather than the application of fertilizers as quick fixes, is at the very heart of organic gardening.

But like all gardeners, organic gardeners have to start somewhere. Your soil may be deficient in certain nutrients. It may not have excellent soil structure. Its pH may be too high, or too low. Unless you've lucked into the perfect soil, you're going to have to work to make it ideal for gardening.

Types of Organic Fertilizer

Dry Organic Fertilizers

Dry organic fertilizers can consist of a single material, such as rock phosphate or kelp (a type of nutrient-rich seaweed), or they can be a blend of many ingredients. Almost all organic fertilizers provide a broad array of nutrients, but blends are specially formulated to provide balanced amounts of nitrogen, potassium, and phosphorus, as well as micronutrients. There are several commercial blends, but you can make your own general-purpose fertilizer by mixing individual amendments.

The most common way to apply dry fertilizer is to broadcast it and then hoe or rake it into the top 4 to 6 inches of soil. You can add small amounts to planting holes or rows as you plant seeds or transplants. Unlike dry synthetic fertilizers, most organic fertilizers are nonburning and will not harm delicate seedling roots. During the growing season, boost plant growth by side-dressing dry fertilizers in crop rows or around the drip line of trees or shrubs. It's best to work side-dressings into the top inch of the soil.

Liquid Organic Fertilizers

Plants can absorb liquid fertilizers through both their roots and through leaf pores. Foliar feeding (that's through the leaves) can supply nutrients when they are lacking or unavailable in the soil, or when roots are stressed. It is especially effective for giving fast-growing plants like vegetables an extra boost during the growing season. Some foliar fertilizers, such as liquid seaweed (kelp), are rich in micronutrients and growth hormones. These foliar sprays also appear to act as catalysts, increasing nutrient uptake by plants. Compost tea and seaweed extract are two common examples of organic foliar fertilizers.

Use liquid fertilizers to give your plants a light nutrient boost or snack every month or even every two weeks during the growing season. With flowering and fruiting plants, foliar sprays are most useful during critical periods (such as after transplanting or during fruit set) or periods of drought or extreme temperatures. For leaf crops, some suppliers recommend biweekly spraying.

When using liquid fertilizers, always follow label instructions for proper dilution and application methods. You can use a surfactant, such as coconut oil or a mild soap (¼ teaspoon per gallon of spray), to ensure better coverage of the leaves. Otherwise, the spray may bead up on the foliage and you won't get the maximum benefit. Measure the surfactant carefully; if you use too much, it may damage plants. A slightly acid spray mixture is most effective, so check your spray's pH. Use small amounts of vinegar to lower pH and baking soda to raise it. Aim for a pH of 6.0 to 6.5.

To apply, use any sprayer or mister, from hand-trigger units to knapsack sprayers, set on the finest spray setting (but never use one that has been used to apply herbicides). The best times to spray are early morning and early evening when the liquids will be absorbed most quickly and won't burn foliage. Choose a day when no rain is forecast and temperatures aren't extreme. Spray until the liquid drips off the leaves. Concentrate the spray on leaf undersides, where leaf pores are more likely to be open. You can also water in liquid fertilizers around the root zone. A drip irrigation system can carry liquid fertilizers to your plants. Kelp is good product for this use (fish emulsion can clog the irrigation emitters).

Growth Enhancers

Growth enhancers are materials that help plants absorb nutrients more effectively from the soil. The most common growth enhancer is kelp (a type of seaweed), which has been used by farmers for centuries. Kelp is sold as a dried meal or as an extract of the meal in liquid or powdered form. It is totally safe and provides some 60 trace elements that plants need in very small quantities. It also contains growth-promoting hormones and enzymes.

To apply, follow the directions for spraying liquid fertilizers when applying growth enhancers as a foliar spray. You can also apply kelp extract or meal directly to the soil; soil application will stimulate soil bacteria, which in turn increases fertility through humus formation, aeration, and moisture retention. Apply 1 to 2 pounds of kelp meal per 100 square feet of garden each spring once a month for the first 4 to 5 months of the growing season.

Or, if you can get your hands on fresh seaweed, rinse it to remove the sea salt and spread it over the soil surface in your garden as a mulch, or compost it. Seaweed decays readily because it contains little cellulose.

Chemical vs. Organic Fertilizer

Many organic materials serve as both fertilizers and soil conditioners—they feed both soils and plants. This is one of the most important differences between a chemical approach and an organic approach toward soil care and fertilizing. Soluble chemical fertilizers contain mineral salts that plant roots can absorb quickly. However, these salts do not provide a food source for soil microorganisms and earthworms, and they will even repel earthworms because they acidify the soil. Over time, soils treated only with synthetic chemical fertilizers lose organic matter and the all-important living organisms that help to build a quality soil. As soil structure declines and water-holding capacity diminishes, more and more of the chemical fertilizer applied will leach through the soil. In turn, it will take ever-increasing amounts of chemicals to stimulate plant growth. When you use organic fertilizers, you avoid throwing your soil into this kind of crisis condition.

What's more, the manufacturing process of most chemical fertilizers depends on non-renewable resources, such as coal and natural gas. Others are made by treating rock minerals with acids to make them more soluble. Fortunately, there are more and more truly organic fertilizers coming on the market. These products are made from natural plant and animal materials or from mined rock minerals. However, the national standards that define and distinguish organic fertilizers from chemical fertilizers are complicated, so it's hard to be sure that a commercial fertilizer product labeled "organic" truly contains only safe, natural ingredients. Look for products labeled "natural organic," "slow release," and "low analysis." Be wary of products labeled organic that have an NPK (nitrogen-phosphorus-potassium) ratio that adds up to more than 15. Ask a reputable garden center owner to recommend fertilizer brands that meet organic standards, or go the DIY route and make your own organic fertilizer.

How to use Organic Fertilizers

If you're a gardener who's making the switch from chemical to organic fertilizers, you may be afraid that using organic materials will be more complicated and less convenient than using premixed chemical fertilizers. Not so! Organic fertilizer blends can be just as convenient and effective as blended synthetic fertilizers. You don't need to custom feed your plants organically unless it's an activity you enjoy. So while some experts will spread a little blood meal around their tomatoes at planting, and then some bonemeal just when the blossoms are about to pop, most gardeners will be satisfied to make one or two applications of general-purpose organic fertilizer throughout the garden.

Make Fertilizer Faster with the Ultimate Compost Bin

Convenient products like dehydrated organic cow manure pellets, liquid seaweed, and fish emulsion make it easy to fertilize houseplants and containers, too. (Don't use fish emulsion indoors, though, because of its strong odor. Save it for your outdoor containers and garden plants).

If you want to try a plant-specific approach to fertilizing, you can use a variety of specialty organic fertilizers, such as mixes designed for tomatoes, roses, transplants, lawns, heavy bloom production, and even container gardens. You can also make custom mixes to address your plants' specific needs. For example, you can use bat and bird guano, composted chicken manure, blood meal, chicken feather meal, or fish meal as nitrogen sources. Bonemeal is a good source of phosphorus, and kelp or greensand are organic sources of potassium.

Poultry Litter

In agriculture, poultry litter or broiler litter is a mixture of poultry excreta, spilled feed, feathers, and material used as bedding in poultry operations. This term is also used to refer to unused bedding materials. Poultry litter is used in confinement buildings used for raising broilers, turkeys and other birds. Common bedding materials include wood shavings, sawdust, peanut hulls, shredded sugar cane, straw, and other dry, absorbent, low-cost organic materials. Sand is also occasionally used as bedding. The bedding materials help absorb moisture, limiting the production of ammonia and harmful pathogens. The materials used for bedding can also have a significant impact on carcass quality and bird performance.

There are specific practices that must be followed to properly maintain the litter and maximize the health and productivity of the flocks raised on it. Many factors must be considered in successful litter management including time of the year, depth of the litter, floor space per bird, feeding practices, disease, the kind of floor, ventilation, watering devices, litter amendments, and even the potential fertilizer value of the litter after it is removed from the house. Most poultry are grown on dirt floors with some type of bedding material. Concrete floors and some specialized raised flooring are used at some facilities. In many areas of the United States, shavings from pine or other soft woods have historically been the bedding of choice for poultry production. Regionally, other materials have been the bedding material of choice due to regional cost and availability, such as rice hulls in the lower Mississippi River poultry production areas of Arkansas and Mississippi.

Bedding Materials

Growers consider a number of factors when determining which material to use as bedding in their facilities, with cost and availability being a major consideration. Bedding materials generally needs to be very absorbent, and must have a reasonable drying time. Many paper products, for instance, absorb moisture well but do not dry out appropriately. The material should also have a useful purpose once it has been used as a bedding material. Without a useful purpose for the used litter, poultry growers would need to dispose of unmanageable quantities of old litter. Large accumulations of litter stored unused for long periods of time are not ecologically acceptable even on a small scale, and would be non-sustainable from an industrial perspective.

Poultry bedding materials also have to be reasonably available. Some materials may meet industry goals once under the birds but if it is difficult to obtain, it will not find favor as a poultry litter. Finally, if a material is not cost competitive with current materials utilized, it will also not be used as a litter material. However, if the new material has increased value once removed from the poultry house compared to current litters or if the current litter material itself becomes difficult to obtain or the quality is decreases, poultry growers may decide to use the new litter material.

Bedding material must not be toxic to poultry or to poultry growers. It should not be excessively favorable for the growth of the litter beetle, a major pest. The effect on other livestock, pets, wildlife, and even plants must also be considered. Poultry can consume as much as 4% of their diet as litter, therefore any bedding material must not contain contaminants, such as pesticides or metals. Consumption by the birds due to litter eating or other bird behavior could affect production and potentially cause the meat or rendered products to become unusable. Pine shavings has been the bedding of choice because of performance, availability, and cost.

Management Practices

Moisture

The heating and ventilation systems in a poultry house must be continuously monitored to keep the moisture content of the litter controlled so that the litter remains friable (easily broken up or crumbly). If the litter becomes too wet and the litter is allowed to become "sealed", then the birds will be living on a damp, slippery and sticky surface. This sealed litter is what is referred to as being "caked." In this condition, the litter is simply saturated with water and is unable to dry out. A severe problem with litter moisture will result if large areas of the house floor surface are caked. The more common issue, however, is having localized areas of caking near leaky watering cups, nipples, troughs or roofs. Watery droppings caused by nutrition and/or infectious agents can also be a cause of excessive moisture in poultry litter.

If litter is not kept at an acceptable moisture level, very high bacterial loads and unsanitary growing conditions may result producing odors (including ammonia), insect problems (particularly flies), soiled feathers, footpad lesions and breast bruises or blisters. This can affect the health and mortality rate of the flock, and could result in quality issues when birds reared under such poor conditions reach the processing plant. In a well-managed broiler house, litter moisture normally averages between 25 and 35 percent. Litter that is managed correctly with the moisture content kept within the acceptable range can be reused if no disease or other production problems occur. On the other hand, caked litter must be removed between flocks and replaced with new litter.

Litter Re-utilization

Some broiler producers are simply removing cake and excess litter after house washing and then placing broilers on old litter for an extended number of flocks. Their expectation is that total clean out is not needed unless there some disease or other bio-security issues. However, producers doing this should be aware that total disinfection under these conditions is probably not possible.

Re-utilization of at least some fraction of used litter as a supplement for fresh wood shavings bedding in broiler houses has been found to not significantly increase pathogens and indicator microorganisms in litter compared to using fresh wood shavings. No consistent significant differences have been found regarding flock performance when comparing houses using fresh litter with houses re-utilizing litter.

A major issue with re-utilization of previously used litter is the generation of ammonia. Ammonia is produced by microbial breakdown of fecal material in the litter. It is well documented in the literature that higher moisture levels result in higher ammonia production. The caked portion of the litter is very high in moisture and nitrogen and should be removed from the house to reduce ammonia generation and provide optimal air quality for chicks during the brooding period. Add litter treatments to reduced ammonia generation. Controlling ammonia with a litter treatment can save money on energy costs by reducing the amount of air exchange required to maintain adequate air quality.

Litter Amendments

High ammonia levels in poultry houses can result in poor bird performance and health and a loss of profits to the grower and integrator. When broilers and turkeys are raised on litter, amendments can be used to reduce ammonia levels in the houses and improve productivity. Uric acid and organic nitrogen (N) in the bird excreta and spilled feed are converted to ammonium (NH_4+) by the microbes in the litter. Ammonium, a plant-available N form, can bind to litter and also dissolve in water. Depending on the moisture content, temperature, and acidity of the litter, a portion of the ammonium will be converted into ammonia (NH_3). Ammonia production is favored by high temperature and high pH (i.e., alkaline conditions). Ammonia is a pungent gas that irritates the eyes and respiratory system and can reduce resistance to infection in poultry. At high-enough concentrations, ammonia will reduce feed efficiency and growth while increasing mortality and carcass condemnations. The result is economic loss to the grower and integrator. Because chicks are more susceptible to the negative effects of ammonia, placing broods in houses with high levels of builtup litter is particularly harmful. Also, the high temperatures required during brooding increase ammonia levels, and moist litter (due to leaky drinkers or high water tables) and insufficient winter ventilation contribute to high ammonia levels as well. In these situations, some growers rely mainly on ventilation to reduce ammonia in the houses. However, ammonia loss from the litter reduces its fertilizer value, and venting ammonia into the environment can cause health and environmental problems.

There are several types of litter amendments available to manage ammonia the most common being acidifiers, and various microbial and enzymatic treatments.

Acidifiers

This type of amendment creates acidic conditions (pH less than 7) in the litter, resulting in more of the ammoniacal nitrogen being temporarily retained as ammonium rather than ammonia. Ammonium is a highly reactive ion that bonds with sulfates, nitrates and phosphates to form ammonium salts that improve the nutrient value of litter when land applied as fertilizer. The acidity also creates unfavorable environment for urolytic bacteria reducing the production of enzymes that contribute to ammonia formation, resulting in reduced ammonia production. Urolytic bacteria have a pH optimum of approximately 8.3 and litter amendments lower the litter surface pH to below 4.0 for a short period of time, usually between 3–5 days depending on the litter amendment. There are several different types of acidifiers, such as alum, acidified liquid alum, sodium bisulfate, ferric sulfate, and sulfuric acid, that have been used by the poultry industry. These products vary in effectiveness as the pH is raised by the activity of the poultry within a couple of weeks. The combination of using litter amendments and poultry house ventilation provides a healthy and comfortable rearing environment. Controlling volatilized ammonia in poultry and livestock rearing environments is critical to maintaining a high level of animal health, well-being and efficient live performance. Most of these products are regulated by DOT and HAZMAT and physical properties for the different products rang from mild irritant to corrosive. Most litter acidifiers are corrosive but when applied according to manufacturers instructions are safe and effective.

Other Amendments

There have been a number of other substances used for ammonia control. A study in Finland found that peat, which is high in humic acid, when used as poultry litter it was quite effective in controlling ammonia. A number of products have also appeared on the market using de-nitrifying or nitrogen-fixing bacteria.

Windrowing

One common practice is windrow composting. This is a deep stacking of litter, usually by plowing the litter into long rows the length of the poultry house. This is an incomplete composting process, and can eliminate harmful pathogens such as E. coli and Salmonella providing that the internal stack temperature reaches 140 to 160 °F. Re-spreading the stacked litter and allowing it to dry would be expected to decrease ammonia and extend litter life.

Disposal and Re-use

Broilers have on average a 47-day growout period, during which the typical broiler chicken will generate about two pounds of litter, if you add the manure and bedding materials. Actual manure generation will be lower because it is only a fractional component of litter. This translates to an average of about 0.7 ounce per day per bird, varying considerably over the life of the bird. This means that a single broiler house, which can contain well over 20,000 birds can generate over 40,000 lbs of litter per flock.

Historically, applications for used poultry litter have included, and still include, use as feed for cattle in the commercial beef industry, land application as a fertilizer for crops or pastures, or occasionally as potting material for the greenhouse and plant container industries. Recently there has been an upsurge in the use of poultry litter as a bio-fuel source for electrical cogeneration and gasification.

Use as Fertilizer

Poultry litter's traditional use is as fertilizer. As with other manures, the fertilizing value of poultry litter is excellent, but it is less concentrated than chemical fertilizers, giving it a relatively low value per ton. This makes it uneconomical to ship long distances, and it tends to lose its nitrogen value fairly quickly. Extracting its value requires that it be used on nearby farms. This limits its resale value in regions where there are more poultry farms than suitable nearby farmland. Poultry litter is also a source of nutrients to the crops, it contains high level of nutrients such as N, P and C.

Use as Cattle Feed

Traditionally used as fertilizer, it is now also used as a livestock feed as a cost-saving measure compared with other feedstock materials, particularly for beef animals.

The use of poultry litter as food for beef cattle is legal in the United States. Prior to 1967, the use of poultry litter as cattle feed was unregulated but that year the FDA issued a policy statement that poultry litter offered in interstate commerce as animal feed was adulterated, effectively banning the practice. In 1980, FDA reversed this policy and passed regulation of litter to the states. In December 2003, in response to the detection of bovine spongiform encephalopathy (mad cow disease) in a cow in the state of Washington, the FDA announced plans to put in place a poultry litter ban. Because poultry litter can contain recycled cattle proteins as either spilled feed or feed that has passed through the avian gut, the FDA was concerned that feeding litter would be a pathway for spreading mad cow disease. In 2004, FDA decided to take a more comprehensive approach to BSE that would remove the most infectious proteins from all animal feeds. The FDA decided at this point that a litter ban was unnecessary in part based on comments by the North American Rendering Industry. In 2005, the FDA published a proposed rule that did not include a litter ban and in 2008 the final rule did not include the ban either.

Use as Fuel

There are currently several electrical generating plants in the UK, and recently in the US, that are utilizing poultry and turkey litter as their primary fuel. The first three (and the world's first three of these plants) were developed by Fibrowatt Ltd in the UK, founded by Simon Fraser, who was appointed an OBE for his contribution to renewable energy. These are: Thetford (38.5 MWe), Eye (12.7 MWe) and Glanford (13.5 MWe - now switched to burning meat and bonemeal). The fourth, Westfield (9.8 MWe), was developed by Energy Power Resources, which now owns all four. Simon Fraser's son and partner, Rupert Fraser, went on to develop the first US plant.

On a smaller scale, poultry litter is used in Ireland as a biomass energy source. This system uses the poultry litter as a fuel to heat the broiler houses for the next batch of poultry being grown thus removing the need for LPG gas or other fossil fuels.

Some companies are also developing gasification technologies to utilize poultry litter as a fuel for electrical and heating applications, along with producing valuable by-products including activated carbons and fertilizers.

Organic Manure

Manures may be defined as materials which are organic in origin, bulky and concentrated in nature and capable of supplying plant nutrients and improving soil physical environment having no definite chemical composition with low analytical value produced from animal, plant and other organic wastes and by products.

Organic manures are included well rotten farm yard manure (FYM), compost, green manures etc. Generally farm yard manures and composts are the decomposed products of agricultural by-products (animals and crops). Whereas green manures may be defined as materials which are un-decomposed green plant tissues susceptible to decomposition in the soil after incorporation.

Classification of Organic Manures

We know that organic manures are of different types i.e. mainly bulky and concentrated in nature.

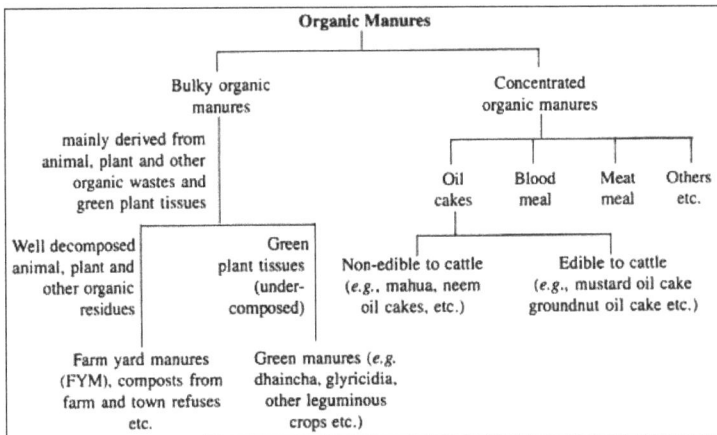

A simple classification scheme of organic manures.

Bulky Organic Manures

Bulky organic manures generally contain fewer amounts of plant nutrients as compared to concentrated organic manures. The concentrated organic manures are mainly derived from raw materials of animal or plant origin.

The amount of nutrients content varies with the nature and kind of oil cakes. No definite composition of NPK and other micro-nutrients can be given. However, all oil cakes either edible or non-edible contains differential amount of N, P and K etc.

Farm Yard Manure

The term farm yard manure refers to the well-decomposed mixture of dung, urine, farm litter (bedding material) and left over or used up materials from roughages or fodder fed to the cattle. The FYM collected daily from the cattle shed consisting of raw dung and part of the urine absorbed in the refuse.

Newly collected and stored FYM is fresh as against well decomposed FYM which has been stored for a sufficient period of time to allow its decomposition to completion.

Farm yard manure consists of two components solid phase, dung and liquid phase, urine. On an average, the animals give out three parts by weight of dung and one part by weight of urine. However, this ratio of dung and urine varies with the kind of animals.

Table: Average amount of moisture and nutrient content of different animals.

Animals	Dung : Urine Ratio	Moisture (%)	'Nutrients' (kg/t) Nitrogen (N)	Phosphorus (P_2O_5)	Potasium (K_2O)
Dairy Cattle	80 : 20	85	4.53	1.22	3.40
Feeder Cattle	80 : 20	85	5.40	2.13	3.22
Poultry	100 : 0	62	13.54	6.48	3.17
Swine	60 : 40	85	5.84	3.22	4.94
Sheep	67 : 33	66	10.42	3.17	9.83
Horse	80 : 20	66	66.75	2.03	5.98

Horses, cows and bullocks give out more dung and less urine than that of sheep, goats and pigs. An average nutrient and moisture content of different kinds of animals is given.

Urine of all animals contains more amounts of N and P as compared to dung. The urine of cows, bullocks and horses contain practically nil or traces of P. Similarly, the dung of all animals except pig is low in P.

Concentrated Organic Manures

Concentrated organic manure may be defined as a material of organic origin derived from raw materials of animal or plant, without bulky in nature having no definite composition of plant nutrients.

Some most common such organic manures are oil cakes edible to cattle (e.g. mustard oil cake, groundnut oil cake, till oil cake etc.) and non-edible to cattle (e.g. neem oil cake, mahua oil cakes etc.); blood-meal, fish manure, bone meal etc.

Since the sources of availability of concentrated organic manures are different and hence they do not contain definite amount of nutrient elements. However, concentrated organic manures are easy to handle and have relatively higher plant nutritive value as compared to bulky organic manures. Besides these, it is quick-acting organic manure when incorporated into the soil.

Characteristics of Organic Manures

All these manures are bulky in nature as well as concentrated nature and supply:

1. Plant nutrients in small quantities,

2. Organic matter in large amounts.

Since it contains two components (plant nutrients and organic matter), when it is applied into the soil it will act as follows:

1. Organic manures supply primary, secondary and micro-nutrients to plants which are liberated in an available forms during the process of mineralization carried out by different micro-organisms.

2. Organic manures also supply organic matter to the soil and hence improve the physical condition of the soil like soil structure, aeration, water holding capacity etc.

3. It also stimulates the activity of different soil micro-organisms through the supply of energy.

4. It improves the buffering and exchange capacities of soil and also influences the solubility of soil minerals as well as mineral nutrients in soil.

5. It also forms chelates which also help for the nutrition of plants.

6. It also regulates the thermal regimes of the soil.

Factors Affecting the Composition of Organic Manures

There are various factors which can affect the composition as follows:

1. Origin of Manure:

   Sheep and poultry manures are somewhat richer in plant nutrients than cow, horse and pig manures.

2. Types of Food Consumed by Animals:

   This is one of the most important factors that determine the manure quality. As for example, the richer the food in proteins, the richer will be the manure in nitrogen.

3. Age and Condition of Animals:

   The manure of young animals is not so rich like that of matured animals because young animals retain more nutrients for their growth than that of old or matured ones.

4. Species of Animals:

    The composition of nutrient contents varies with the ruminant and non-ruminant animals.

5. Nature and Amount of Litter:

    The composition of FYM varies with the nature and amount of litter used for animals (e.g. paddy straws, wheat straws etc.)

6. Function of the Animal:

    Animals producing milk and wool absorb large amount of nutrients from their food than that of working draft animals. Therefore, manure from bullocks generally contains more nutrients as compared to milch cows.

7. Handling and Storage of Manures:

    Loss of potash occurs if any drainage is allowed to escape from the manure heap. Therefore, improper handling and storage leads to losses of plant nutrients from the manures.

Reactions of Organic Manures in Soils

It is found that both bulky and concentrated organic manures contain some amount of plant nutrients including macro- and micro-nutrients, of which organic nitrogen content is likely to be dominant. The organic forms of soil nitrogen occur as consolidated amino acids, proteins, amino sugars etc.

When Organic manures like FYM, composts, oil cakes, green manures etc. are added to the soil, the microbial attack to these materials takes place and results complete disappearance of the organic protein with the remainder of the nitrogen being changed into inorganic form of nitrogen through the process of mineralization.

Proteins $\rightarrow R - NH_2 + CO_2$ + energy + other products
(present in organic manures)

$$R - NH_2 + HOH \rightarrow NH_3 + R - OH + energy$$

$$\searrow + H2O$$

$$NH_4^+ + OH^-$$

(release of ammonium in the soil)

The released ammonium (NH_4^+) is subject to following changes:

1. It may be converted to nitrites and nitrates through the process of nitrification carried out by micro-organisms.

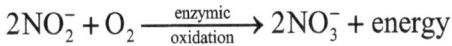

$$2NH_4^+ + 3O_2 \xrightarrow[\text{oxidation}]{\text{enzymic}} 2NO_2^- + 2H_2O + 4H^+ + \text{energy}$$

$$2NO_2^- + O_2 \xrightarrow[\text{oxidation}]{\text{enzymic}} 2NO_3^- + \text{energy}$$

2. It may be absorbed directly by the plants.

3. It may be utilized by hetero-trophic organisms in further decomposing organic carbon residues.

4. It may be fixed in the lattice of certain expanding-type clay minerals.

5. It could be slowly released back to the atmosphere as elemental nitrogen.

Mineralization and immobilization of nitrogen or any other nutrient elements occur continuously in microbial metabolism and the magnitude and direction of the net effect are greatly influenced by the nature and amount of organic manures added.

Normally organic manures are applied to the soil as a source of fertilizer nitrogen should contain about 1.5 to 2.0 per cent of the dry weight of the manures in order to meet the needs of the soil microorganisms, otherwise little or no nitrogen will be released for the use of plants.

The carbon nitrogen ratio (C:N ratio) in the organic manures remaining in the soil after consuming by the soil micro-organisms is approximately 10: 1.

Therefore, different organic manures containing variety of organically bound nutrients like P, S and other micro-nutrients etc. are subject to transformation in soils similar to that of mineralization and immobilization processes of nitrogen and releases inorganic forms of nutrients in soils which become available to plants.

Green Manure

In agriculture, green manure is created by leaving uprooted or sown crop parts to wither on a field so that they serve as a mulch and soil amendment. The plants used for green manure are often cover crops grown primarily for this purpose. Typically, they are ploughed under and incorporated into the soil while green or shortly after flowering. Green manure is commonly associated with organic farming and can play an important role in sustainable annual cropping systems.

Functions

Green manures usually perform multiple functions that include soil improvement and soil protection:

1. Leguminous green manures such as clover and vetch contain nitrogen-fixing symbiotic bacteria in root nodules that fix atmospheric nitrogen in a form that

plants can use. This performs the vital function of fertilization. If desired, animalmanures may also be added.

Depending on the species of cover crop grown, the amount of nitrogen released into the soil lies between 40 and 200 pounds per acre. With green manure use, the amount of nitrogen that is available to the succeeding crop is usually in the range of 40-60% of the total amount of nitrogen that is contained within the green manure crop.

Average biomass yields and nitrogen yields of several legumes by crop:	Biomass tons acre^{-1}	N lbs acre^{-1}
Sweet clover	1.75	120
Berseem clover	1.10	70
Crimson clover	1.40	100
Hairy vetch	1.75	110

2. Green manure acts mainly as soil-acidifying matter to decrease the alkalinity/pH of alkali soils by generating humic acidand acetic acid.

3. Incorporation of cover crops into the soil allows the nutrients held within the green manure to be released and made available to the succeeding crops. This results immediately from an increase in abundance of soil microorganisms from the degradation of plant material that aid in the decomposition of this fresh material. This additional decomposition also allows for the re-incorporation of nutrients that are found in the soil in a particular form such as nitrogen (N), potassium(K), phosphorus (P), calcium (Ca), magnesium (Mg), and sulfur (S).

4. Microbial activity from incorporation of cover crops into the soil leads to the formation of mycelium and viscous materials which benefit the health of the soil by increasing its soil structure (i.e. by aggregation).

The increased percentage of organic matter (biomass) improves water infiltration and retention, aeration, and other soil characteristics. The soil is more easily turned or tilled than non-aggregated soil. Further aeration of the soil results from the ability of the root systems of many green manure crops to efficiently penetrate compact soils. The amount of humusfound in the soil also increases with higher rates of decomposition, which is beneficial for the growth of the crop succeeding the green manure crop. Non-leguminous crops are primarily used to increase biomass.

5. The root systems of some varieties of green manure grow deep in the soil and bring up nutrient resources unavailable to shallower-rooted crops.

6. Common cover crop functions of weed suppression. Non-leguminous crops

are primarily used (e.g. buckwheat) The deep rooting properties of many green manure crops make them efficient at suppressing weeds.

7. Some green manure crops, when allowed to flower, provide forage for pollinating insects. Green manure crops also often provide habitat for predatory beneficial insects, which allow for a reduction in the application of insecticides where cover crops are planted.

8. Some green manure crops (e.g. winter wheat and winter rye) can also be used for grazing.

9. Erosion control is often also taken into account when selecting which green manure cover crop to plant.

10. Some green crops reduce plant insect pests and diseases. Verticillium wilt is especially reduced in potato plants.

Incorporation of green manures into a farming system can drastically reduce the need for additional products such as supplemental fertilizers and pesticides.

Limitations to consider in the use of green manure are time, energy, and resources (monetary and natural) required to successfully grow and utilize these cover crops. Consequently, it is important to choose green manure crops based on the growing region and annual precipitation amounts to ensure efficient growth and use of the cover crop(s).

Nutrient Creation

Green manure is broken down into plant nutrient components by heterotrophic bacteria that consumes organic matter. Warmth and moisture contribute to this process, similar to creating compost fertilizer. The plant matter releases large amounts of carbon dioxide and weak acids that react with insoluble soil minerals to release beneficial nutrients. Soils that are high in calcium minerals, for example, can be given green manure to generate a higher phosphate content in the soil, which in turn acts as a fertilizer.

The ratio of carbon to nitrogen in a plant is a crucial factor to consider, since it will impact the nutrient content of the soil and may starve a crop of nitrogen, if the incorrect plants are used to make green manure. The ratio of carbon to nitrogen will differ from species to species, and depending upon the age of the plant. The ratio is referred to as C:N. The value of N is always one, whereas the value of carbon or carbohydrates is expressed in a value of about 10 up to 90; the ratio must be less than 30:1 to prevent the manure bacteria from depleting existing nitrogen in the soil. Rhizobium are soil organisms that interact with green manure to retain atmospheric nitrogen in the soil. Legumes, such as beans, alfalfa, clover and lupines, have root systems rich in rhizobium, often making them the preferred source of green manure material.

Crops

Late-summer and fall green manure crops are oats and rye. Other green manure crops:

1. Alfalfa, which sends roots deep to bring nutrients to the surface,

2. Buckwheat in temperate regions,

3. Cowpea,

4. Clover (e.g. annual sweet clover),

5. Fava beans,

6. Fenugreek,

7. Ferns of the genus *Azolla* have been used as a green manure in southeast Asia,

8. Lupin,

9. Groundnut,

10. Millet,

11. Mustard,

12. *Phacelia tanacetifolia,*

13. Radish such as tillage radish or daikon radish,

14. Sesbania,

15. Sorghum,

16. Soybean,

17. Sudangrass,

18. Sunn hemp, a legume widely grown throughout the tropics and subtropics,

19. Tyfon, a *Brassica* known for a strong tap root that breaks up heavy soils,

20. Velvet bean (*Mucuna pruriens*), common in the southern US during the early part of the 20th century, before being replaced by soybeans, popular today in most tropical countries, especially in Central America, where it is the main green manure used in slash/mulch farming practices.

Phosphate Rich Organic Manure

Phosphate rich organic manure is a type of fertilizer used as an alternative to diammonium phosphate and single super phosphate.

Phosphorus is required by all plants but is limited in soil, creating a problem in agriculture In many areas phosphorus must be added to soil for the extensive plant growth that is desired for crop production. Phosphorus was first added as a fertilizer in the form of single super phosphate (SSP) in the mid-nineteenth century, following research at Rothamsted Experimental Station in England. SSP is non-nitrogen fertiliser containing Phosphate in the form of monocalcium phosphate and Gypsumwhich is best suited for Alkali soils to supplement Phosphate and reduce soil alkalinity.

The world consumes around 140 million tons of high grade rock phosphate mineral today, 90% of which goes into the production of diammonium phosphate (DAP). Excess application of chemical fertilizers in fact reduces the agricultural production as chemicals destroy natural soil flora and fauna. When DAP or SSP is applied to the soil only about 30% of the phosphorus is used by the plants, while the rest is converted to forms which cannot be used by the crops a phenomenon which is known as phosphate problem to soil scientists.

Phosphate Rich Organic Manure is produced by co-composting high-grade (32% P_2O_5 +/- 2%) rock phosphate in very fine size (say 80% finer than 54 microns). Needless to say, the finer the rock phosphate the better is the agronomic efficiency of PROM. Research indicates that this substance may be a more efficient way of adding phosphorus to soil than applying chemical fertilizers. Other benefits of PROM are that it supplies phosphorus to the second crop planted in a treated area as efficiently as the first, and that it can be produced using acidic waste solids recovered from the discharge of biogas plants.

Phosphorus in rock phosphate mineral is mostly in the form of tricalcium phosphate, which is water-insoluble. Phosphorus dissolution in the soil is most favorable at a pH between 5.5 and 7. Ions of aluminum, iron, and manganese prevent phosphorus dissolution by keeping local pH below 5.5, and magnesium and calcium ions prevent the pH from dropping below 7, preventing the release of phosphorus from its stable molecule. Microorganisms produce organic acids, which cause the slow dissolution of phosphorus from rock phosphate dust added to the soil, allowing more phosphorus uptake by the plant roots. Organic manure can prevent ions of other elements from locking phosphorus into insoluble forms. The phosphorus in phosphate enhanced organic manure is water-insoluble, so it does not leach into ground water or enter runoff.

Most phosphate rocks can be used for phosphate rich organic manure. It was previously thought that only those rocks which have citric acid soluble phosphate and those of sedimentary origin could be used. Rocks of volcanic origin can be used as long as they are ground to very fine size.

Organic manure should be properly prepared for use in agriculture, reducing the C:N ratio to 30:1 or lower. Alkaline and acidic soils require different ratios of phosphorus.

PROM is known as a green chemistry phosphatic fertilizer. Addition of natural minerals or synthetic oxides in water-insoluble forms that contain micronutrients such as copper, zinc, and cobalt may improve the efficiency of PROM. Using natural sources of nitrogen, such as *Azolla*, may be more environmentally sound.

Phosphate Rich Organic Manure under FCO

Ministry of Agriculture and Cooperation, Government of India has now approved Phosphate Rich Organic Manure (PROM) and included the same under Fertilizer Control Order (FCO). The approved specifications may be seen from Gazette Notification from the web site of PROM Society.

Advantages and Disadvantages of using Organic Fertilizer

Organic Fertilizer.

Balancing the advantages and disadvantages of organic fertilizers have been the subject of debate for nearly 100 years, with the history of the organic movement dating back to the earliest days of chemical alternatives. For most people, a careful weighing of the options will help you determine which method is best for your lawn or garden.

Benefits of using Organic Options

Until early in the twentieth century, composted manure, kitchen scraps, and other organic wastes represented the only means of improving soil fertility. Organic farming and gardening were not moral or environmental choices, but simply the way of life. Organic fertilizers are an all-encompassing method that build soil structure and composition as they slowly release nutrition to the plants.

Steps for Soil Health

Virtually every aspect of organic gardening revolves around the health of the soil. a Thriving growing substrate revolves around maintaining the delicate symbiotic relationship between soil microorganisms and the nutrients they (and the plants) require. Organic fertilizer contributes to soil health in the following ways:

1. Increased nutrition: Natural soil is rich in organic matter that releases nutrients at a steady rate. Increasing organic matter in agricultural soil improves the soil structure, creating more air space and water retention within the soil. Quality dirt is made up of a mix of particles and substances graded by permeability (how easily air, water and roots can move through it) - granular particles/stone, clay, sand, humus/organic matter. Organic fertilizer assists microorganisms to break down organic matter while allowing a metered release of that nutrition. Organic physical additives like stone dust - and peat or coconut hulls - create good soil with a balanced permeability.

2. Reduced soil erosion: A higher proportion of organic material in the soil will also prevent soil erosion, helping to avoid the dust bowl effect seen in the 1930s. Soil structure is critical to root health. Too permeable soil, like sandy substrates allow too much air and water to move through quickly. Water carries nutrition with it , and quick drainage equals a faster loss of nutrients (this run-off is what damages ecosystems and waterways near industrial agricultural fields). Likewise, clay soils, because of compaction, hold too much water and very little air and "light," dusty earth is prone to the erosion witnessed during the infamous Dust Bowl. Since organic fertilizers are inherently "found" materials - like manure, vegetable scraps, aggregates, seaweed - they build soil structure as they hold and release nutrients to the plants.

3. Healthy Ecosystem: Organic fertilizer is rooted in a complex foundation. Organic fertilizers are not used alone- they are intertwined in a gardening process that is gentler on microorganisms and earthworms living in the soil, creating a healthy ecosystem that is assisted by careful additions of physical (mulch) and nutritional (liquid seaweed) elements. Synthetic or chemical fertilizers create "burn" or bursts of ammonia, phosphorous and nitrogen that give a fast release. These fertilizers, if applied incorrectly, hinder proper fruiting and impede soil ecosystem development. Organic gardeners want to build a healthy, sustainable and environmentally beneficial farming system that becomes self-sustaining with minimal disruption.

Slow and Steady Release of Nutrients

The slow and gradual release of nutrients is listed among both a pro and con of organic fertilizer use. As an advantage, the natural release of elements means that there is a reduced risk of nutrient burn from over-fertilization. This approach also means that applications of soil amendments are required less frequently, reducing operating cost and manual labor. With organic fertilizer, nutrient availability and uptake by plants occur at roughly an equal rate, meaning nutrients are preserved in soil and plant matter rather than leaching away with rainwater. The resulting plant growth occurs at a natural, healthy pace. This tends to produce stronger, more stable plants than those grown at an artificially accelerated rate, theoretically producing improved taste and nutritional value at the same time.

Organic is Economical

Shovel Full of Compost.

Organic fertilizer is potentially a cheaper option than chemical alternatives. If you have a compost or live in a rural area, the only cost is time. Many farmers will sell manure by the truckload or even give it away if you are willing to pick it up. In urban and suburban neighborhoods, a composting unit can be cheap, effective, and unobtrusive. For a nominal upfront investment, even apartment dwellers can have their own organic worm bin composting system to feed a balcony garden.

Better for the Environment

The combined influence of increased organic matter and reduced nutrient leaching means that elements such as nitrogen and phosphorus will end up in your plants' roots instead of the local waterways. Nutrient leaching from agriculture is a major culprit in the development of algae blooms on lakes and ponds. This process, known as eutrophication, disrupts ecosystems and renders water unfit for human use.

Even organic fertilizers can impact the environment if they are not stored or used correctly (think of manure run-off into streams). Since quick release conventional fertilizer products are intended for that fast uptake - the solubility means that drainage carries extra salts off of the growing area at a higher concentration rate. These leached elements enter water supplies as contaminants. Retained chemical fertilizers negatively affect the soil ecosystem creating issues with soil acidification, compaction (or soil crumb), and a systematic destruction of the delicate microflora. Over-use of chemical fertilizers will slowly "kill" off a healthy soil structure.

Always do a soil test before adding any compounds to your garden. Soil testing is easy - simply gather your sample according to the kit's directions and send the sample off. After the initial test, you can retest annually if needed, or every few years.

Drawbacks to Organic Fertilizers

Organic fertilizer holds many advantages over chemical alternatives, but it may not be best in every situation. Conventional fertilizers can be used in a beneficial way - if applied correctly in necessary growing situations. Some potential disadvantages of organics include:

1. Limited nutrient availability: The slow-and-steady approach that makes organic fertilizer perfect for most applications can pose a problem in certain situations. Organic fertilizers are bound into their structures - this is what allows for the slow break-down. The release of nutrients from organic fertilizers can be dependent on both climate and the presence of microorganisms in the soil. Damaged soils may lack the necessary biological conditions for effective composting. Severely nutrient-deprived plants needing a boost might do better initially with a readily available nutrient mixture in a liquid form.

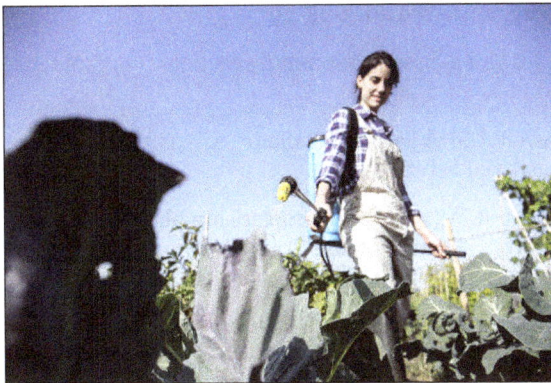

Woman working on farm.

2. Labor-intensive: Organic fertilizers can be bulky, messy materials. Some would argue that working with organic fertilizer is a labor of love, but turning compost piles, moving manure, and spreading solid fertilizer are not for everyone. This also means that applying fertilizer on a large scale can be more difficult, as heavy manure or blood meal granules are less suitable for mechanical spreaders.

3. Potentially pathogenic: Incomplete composting can leave certain pathogens in the organic matter. These pathogens can enter the water system or the food crops, causing human health and environmental problems.

4. Expensive: Commercial organic fertilizers are often more expensive per unit than comparable chemical products.

The negative aspects of manufactured fertilizer can be mitigated and reversed by careful soil testing. Add lime and compost to soils that have been treated with fertilizers. Plant health is negatively affected by the salts and acidification of too much (or too long) use of chemical products, since they can kill off the bacteria and fungal organisms that boost roost health. Mycorrhizal fungi coexist with a plant's root structure (the roots feed the fungi) and the mycorrhizae boost the plant's health by stimulating its immune system and assisting in its nutrient and water uptake.

References

- Organic-fertilizer, definition: maximumyield.com, Retrieved 9 May, 2019

- Organic-fertilizer, ardening, home: goodhousekeeping.com, Retrieved 10 June, 2019

- "biomatnet Item: NNE5-1999-00075 - Power plant based on fluidised bed fired with poultry litter". Biomatnet.org. Retrieved 2013-08-29

- Organic-manures-meaning-classification-and-reactions, organic-manures: Retrieved 11 July, 2019

- Soilmanagementindia.com, Bagley, C.P.; Evans, R.R. (Apr 1995), Broiler litter as a feed or fertilizer in livestock operations, Mississippi State University: Mississippi State University Cooperative Extension Service, ISSN 0886-7488, US9561988

- "biomatnet Item: NNE5-1999-00075 - Power plant based on fluidised bed fired with poultry litter". Biomatnet.org. Retrieved 2013-08-29

- Organic-manures-meaning-classification-and-reactions: organic-manure: soilmanagementindia.com, Retrieved 12 August, 2019

- Piper, C.V.; Pieters A.J. USDA Farmer's Bulletin (ed.). Green Manuring. USDA Farmer's Bulletin. Pp. 1250–1295. Retrieved Feb 2, 2010

- Heathwaite, A.L; Dils, R.M (2000). "Characterising phosphorus loss in surface and subsurface hydrological pathways". Science of the Total Environment. 251-252: 523–538. Doi:10.1016/S0048-9697(00)00393-4. Archived from the original on 2013-02-02

- Commercial-Manufacturing-of-Organic-Fertilizer: organic.lovetoknow.com, Retrieved 13 January, 2019

4

Soil Management in Organic Farming

Soil management includes the application of treatments, operations and practices to protect soil and enhance its performance. Soil management in organic farming focuses on using soil health as the major source for fertilization and pest control. All the techniques and methods related to soil management in organic farming are introduced in this chapter.

Managing soil fertility in organic farming systems requires a different approach from that used in conventional farming systems. Nutrients in synthetic fertilizers are highly soluble, so nutrient availability is quite predictable and nutrients are quickly available to plants. They do not require biological processes to make them available. And they do not enhance the biological health of the soil. In fact, several synthetic fertilizers degrade soil by drying it out or making it acidic or saline. Practices and inputs used by organic producers (as well as many others in sustainable agriculture) promote biologically healthy soils that sustain fertility in ways different from conventional systems. They promote the decomposition of plant and animal residues to make nutrients from them available to plants. Managing soil fertility with biologically-created inputs ensures that organic production is a dynamic biological process. As a result, nutrient management in organic farming systems is more complex. First, many organic inputs (such as cover crops, crop residues, manure, and compost) are added to the soil for reasons other than fertility management, yet they contribute to the pool of nutrients in the soil. Second, most organic materials, including compost and manure, have only a small component of soluble nutrients; most of their nutrients must be transformed through biological processes before they become available to plants. Lastly, most manures and composts do not have a consistent nutrient content as found on bags of synthetic fertilizers. They also contain a ratio of nutrients different from that needed for optimal plant growth. In summary, plant-available nutrients in organic systems include minerals, nutrients that have been mineralized from plant and animal residues, nutrients held in microbial biomass, and nutrients that are being mineralized from decomposing residues. An organic producer needs to account for each of these nutrient sources to determine current and future nutrient availability as well as view nutrient management as but one component of an integrated crop and soil management plan.

Nitrogen Management

Nitrogen is often the most limiting nutrient to efficient and profitable organic crop production. Organic growers are limited to organic sources of nitrogen or those derived from natural processes. Soil organic matter is the backbone of nitrogen supply in organic production, and cultural practices to manipulate nitrogen start with building soil organic matter. Other important potential sources of nitrogen include fixed nitrogen from legumes included as a cover or rotation crop, compost produced from on-farm or off-farm materials, manures from on-farm or off-farm sources, and purchased organic fertilizers.

Organic Sources of Nitrogen

On organic farms, the main source of nitrogen is atmospheric nitrogen fixed by legumes. In a well-designed crop rotation, plow-downs of legume green manures and forage legumes can provide all the nitrogen required to grow cash crops. Nitrogen is kept on the farm by recycling of nutrients from manure and the applications of composted manure. Although raw manure contains more available nitrogen than composted manure, composted manure is preferred for several reasons.

Soil Organic Matter and Humus: Organic growers agree to a soil-building program to maintain or enhance soil organic matter as part of their organic certification requirements, and soil organic matter is one important source of nitrogen for organic production. As previously mentioned, the soil organic matter is divided into active and stable fractions. The stable fraction consists in large part of the humus or heavy fraction and affects overall physical and chemical properties in soil. High C:N ratio crop residues such as straw or corn stalks decompose slowly on incorporation and are more often converted to humus.

Green manure crops: Green manure crops refer specifically to cover crops grown to supply nitrogen and increase soil organic matter. Cover crops are similar to green manures, but are usually grown to conserve topsoil, prevent erosion, improve soil structure, increase organic matter, and capture and hold nutrients during the non-growing season. Examples of green manure crops include grass mixtures and legume plants. Some of the most commonly used are: annual ryegrass, vetch, clover, peas, winter wheat, and alfalfa. As a complement to nitrogen from soil organic matter, a vigorous green manure crop may be the most economical organic source of additional early-season nitrogen for the succeeding crop. In crop production areas with mild climates, green manure crops are widely used in organic farming systems because they grow during the fall, winter, and early spring and can develop substantial biomass.

Manure: Fresh, non-composted manure will generally have higher nitrogen content than composted manure. However, the use of composted manure will contribute more to the organic matter content of the soil. The nitrogen in manure is in two forms: the organic form, which releases slowly; and the inorganic form (ammonium, NH_4^+ and nitrate, NO_3^-), which are immediately available while organic nitrogen that is more stable and slowly released.

Compost: The nitrogen content of composts will vary according to the source material and how it is composted. In general, nitrogen in the form of ammonium (NH_4^+) or nitrate (NO_3^-) is low in composts. The majority of the nitrogen in finished compost (usually over 90%) has been incorporated into organic compounds that are resistant to decomposition. Mineralization rates from compost application are relatively low, and compost is usually a poor short-term source of nitrogen. Rough estimates are that only about 15 percent of the nitrogen in compost will become available the first year following incorporation.

Commercial organic fertilizers: Many types of dry and liquid commercial organic fertilizers are available for use in certified organic crop production. Most of these products are byproducts of fish, livestock, and food and other processing industries. Various single-source and blended organic products are available from different fertilizer suppliers and distributors. Some typical examples of organic fertilizer sources include pelleted poultry manure, seabird guano, pelleted seabird guano, feather meal, and blood meal. Commercial organic nitrogen fertilizers are more concentrated nitrogen sources than compost with improved handling, nitrogen placement, and nitrogen availability. While listed on the National List as a prohibited non-synthetic, sodium nitrate ($NaNO_3$, 16% N) mined from naturally occurring deposits in Chile and Peru can be used in organic production in accordance with its annotation.

Nitrogen Availability

The nitrogen concentration and resulting C:N ratio of the organic materials will affect whether a net release of nitrogen from the organic material will occur during decomposition, known as net nitrogen mineralization, or whether nitrogen from the soil will have to be used up by soil organisms in order to decompose the organic matter, known as net nitrogen immobilization. Soil microbes readily release the excess nitrogen in a plant-available form when organic matter has a low C:N ratio or when nitrogen is in excess.

Nitrogen Synchronization

Synchronizing nitrogen mineralization from soil organic matter, green manure crop residues, and organic amendments to maintain adequate nitrogen availability for crop production is challenging. The rate of nitrogen mineralization from soil organic matter and recently incorporated residues and amendments typically peaks before the crop reaches its maximum rate of nitrogen uptake.

Phosphorus Management

Organic crop production systems seek to improve soil organic matter and biological diversity, which may impact phosphorus cycling and phosphorus uptake by crops. Increases in organic matter will be accompanied by an increase in the organic phosphorus pool. Furthermore, management of green manure crops can increase the availability of soil phosphor us pool (both organic and inorganic) by stimulating microbial activity and release of root exudates.

Organic Sources of Phosphorus

Phosphorus sources approved for use in organic agriculture have diverse properties that affect phosphorus availability and management. Common phosphorus sources include soil organic matter, green manure crops, manure, compost, and rock phosphate, all of which are frequently used in growing crops.

Soil organic matter can be an important source of phosphorus for crops. Soil organic matter contains a variety of organic phosphorus compounds, such as inositol phosphate, nucleic acid, and phospholipid. These compounds must be first converted (i.e., mineralized) to inorganic phosphate by soil enzymes before being used for plant growth. These phosphatase enzymes are produced by soil microorganisms, mycorrhizal fungi, or excreted by the plant root.

Some green manure crops accumulate high levels of phosphorus and are thought to increase phosphorus availability to subsequent crops by returning it to the soil in organic form. Some green manure crops can be excellent hosts for mycorrhizal fungi, which may allow a greater exploitation of the soil phosphorus reserves. In this symbiotic relationship, the plant root provides the energy (carbohydrate) for the fungi in exchange for improved nutrient uptake and other plant root benefits.

Manure and Compost: Phosphorus occurs in livestock manure and in a combination of inorganic and organic forms. In general, 45 to 70 percent of manure phosphorus is inorganic. Unlike nitrogen, phosphorus is conserved in the composting process and, depending on the composting process, the water-soluble phosphorus of mature compost may not be different from that of the original manure source. Organic phosphorus constitutes the rest of total phosphorus. Essentially, all inorganic phosphorus is in the orthophosphate form, which is the form taken up by growing plants.

Colloidal phosphate, also called soft rock phosphate, consists of clay particles surrounded by natural phosphate. Total phosphate is around 20 percent and available phosphate about two to three percent. It also contains about 25 percent lime and other trace minerals. Colloidal phosphate is often added directly to livestock manure, where the manure acids dissolve much of the total phosphate and the phosphate stabilizes the nitrogen in the manure.

Rock phosphate from apatite ore has not been acidulated or otherwise chemically treated. Rock phosphates are usually derived from ancient marine deposits. Synthetic fertilizers such as super- or triple superphosphate are made by reaction with sulfuric or phosphoric acid to increase the solubility of the phosphorus in rock phosphates. Organic farming, which does not permit the use of such acidulation phosphate fertilizers, seeks to convert the phosphorus in rock phosphate to crop-available forms by using soil biological processes and natural soil weathering.

Bone meal, prepared by grinding animal bones, is one of the earliest phosphorus sources used in agriculture. Most commercially available bone meal is "steamed" to remove any raw animal tissue. The primary phosphorus mineral in bone material is "calcium-deficient hydroxyapatite," which is more soluble than rock phosphate, but much less soluble than conventional phosphorus fertilizers.

Guano is most commonly used as a source of nitrogen for plants, but some guano materials are also relatively enriched in phosphorus. Guano is mined from aged deposits of bird or bat excrement in low rainfall environments. The drying and aging process changes the chemistry of the phosphorus compared with fresh manure. Struvite (magnesium ammonium phosphate) is a major, phosphorus mineral found in guano, dissolving slowly in soil. The limited supply and high cost of guano generally restricts its use to small-scale applications.

Mycorrhizal Fungi

Mycorrhizae (my-cor-ry-zee) fungi are special soil fungus that forms a symbiotic mutualistic relationship with plant roots. In this association both organisms benefit, the fungus takes over the role of the plant's root hairs and acts as an extension of the root system. The fungus receives carbohydrates (sugars) and growth factors from the plant, which in turn receives many benefits, including increased nutrient absorption. When a large, vigorous network of mycorrhizae hyphae is associated with a plant's roots, it exponentially expands the "reach" and surface area of those roots, giving the plant greater access to the nutrients, especially phosphorous and other minor nutrients the soil has to offer. This symbiosis is, of course, great for plants, because the extra nutrients can fuel better growth and increase resistance to drought and disease. The value of mycorrhizal fungi for supplying phosphorus for crops is most apparent in low-phosphorus soils.

Potassium Management

Most soils hold large reserves of potassium in the primary and secondary soil minerals, but only a small fraction of the total potassium in soil is immediately available for plant uptake. The potassium in the soil solution and the exchangeable potassium held on the soil colloids are readily available for plant uptake. As the exchangeable potassium is depleted, it can be replenished by the reserves of nonexchangeable potassium, but this is a slow process that is generally not rapid enough to satisfy crop requirements.

Organic Sources of Potassium

Regular applications of soluble potassium, regardless of the source, will increase the concentration of potassium in the soil solution and the proportion of potassium on the cation exchange sites. All of the commonly used soluble potassium sources (including manures, composts, and green manures) contain this nutrient in the simple cationic (K^+) form. Most soluble inorganic fertilizers and organic manures are virtually interchangeable as sources of potassium for plant nutrition. When using readily available forms of potassium, the overall goal of replacing the harvested potassium is generally more important than minor differences in the behavior of the potassium source.

Soil Organic Matter: Potassium is required by plants in amounts second only to nitrogen. Unlike nitrogen and phosphorus, potassium is not organically combined in soil organic matter. Different potassium-containing minerals, such as micas and feldspars, therefore, are the principal sources of potassium in soils.

Manure and Compost: Manure and compost are usually very good sources of available potassium. For most manures and composts, it is generally assumed that the potassium is soluble and that these can be applied on an equal basis with fertilizer potassium recommendations. However, overuse of manure or compost can lead to excessive levels of exchangeable potassium that can interfere with the uptake of calcium and magnesium by crops.

Potassium sulfate: There are two forms of potassium sulfate on the market. One is derived by reacting sulfuric acid with potassium chloride. It is a good fertilizer, but not acceptable in certified organic production. The other is derived from natural sources and it is allowed for organic crop production.

Langbeinite: Langbeinite is listed by the Organic Materials Research Institute (OMRI) as allowable in certified organic production if it is used in the raw, crushed form without any further refinement or purification. Several excellent sources of this approved product are available for use with organic crop production.

Glauconite: Glauconite is a clay-type mineral, commonly sold as greensand, which is listed by OMRI as allowed for organic production. Total potassium oxide (K_2O) content of greensand is around seven percent, but most of the potash is unavailable. The very slow potassium release rate of greensand is touted to minimize the possibility of plant damage by fertilizer "burn," while the mineral's moisture retention may aid soil conditioning.

Management of Micronutrients

Micronutrients essential to plants include calcium sulfur, iron, manganese, boron, copper, zinc, molybdenum, and nickel. In a biological active soil with good CEC and balanced pH, micronutrient deficiencies are rare. Micronutrient deficiencies are most common in sandy soils, and in soils with very low levels of organic matter.

Organic Sources of Micronutrients

Cobalt, copper, iron, manganese, molybdenum, selenium, and zinc—can be applied to correct a deficiency provided if they are from sulfate, carbonate, oxide, or silicate sources. Nitrate and chloride forms of these micronutrients are explicitly prohibited. Synthetic soluble sources of boron can also be applied.

Chelating Agents

Chelating agents are compounds to which an element in its ionic form can be attached. Micronutrients can be made more available to plants by chelation with various compounds. Naturally occurring chelating agents such as citric acid may be used. Synthetic chelating agents on the National List such as lignosulfonic acid and its salts; and humic acids are more commonly used.

Nutrient Management in Organic Farming

The management of nutrients in organic farming systems presents a formidable challenge, as the use of inorganic fertilisers is not permitted. Therefore organic farmers must optimise a range of soil, crop, rotation and manure managements to ensure a nutrient supply which will guarantee optimum crop yields and minimise losses to the environment. To achieve this objective, an appreciation of the nutrient cycles in farming systems is essential.

With organic agriculture only having been a small proportion of total agriculture to date, and with its role as a counterpoint to mainstream farming, the volume of scientific research into organic agricultural techniques is disproportionably small.

There is little doubt that when great care is taken with the storage and spreading of animal manures on the farm, soil fertility can be protected to a considerable extent. Areas used for silage production must get priority when spreading animal manures and ideally the grazing and silage areas should be rotated annually. However, with the best will in the world, there will still be some loss of nutrients. In order for organic agriculture to be sustainable in the long term, these losses must be replaced.

Healthy Soil: The Foundation of Organic Agriculture

Optimising soil health is a key foundation of organic agriculture. Particular emphasis is placed on maintaining good levels of soil biological activity and organic matter coupled with balanced/optimum nutrient levels. Organic agriculture aims to 'feed the soil to feed the plant' by maintaining soil biology and nutrients at optimum levels throughout the rotation rather than the non-organic approach of applying nutrients to feed the current crop to maximise yield. Organics therefore takes a long term, whole farm/systems approach to nutrient management based on regular soil tests and nutrient budgets to determine when soil nutrients must be replaced. As for non-organic farming, the results and fertiliser recommendations of soil tests are tailored to a fields cropping history and soil type to give specific recommendations for each field.

Organic agricultural practices contain something of a dichotomy regarding nutrient management. One view is the 'law of return' where it is considered essential that any nutrients removed in crops or livestock must be returned to maintain fertility i.e., a balanced nutrient budget. The other view considers the farm to be a 'closed system' for nutrients and that they should be carefully recycled within the farm and the need to 'import' nutrients is considered a system failure. However, the latter approach does not consider the removal of nutrients in crops and livestock or losses from leaching or to the atmosphere. The scientific evidence is now conclusive, that both must be combined: efficient cycling of nutrients around the farm must be coupled with the law of return (balanced budgets). This means that it is absolutely essential that all nutrients removed from the land in crops and livestock sold 'through the farm gate' must be replaced. In addition, while livestock can be highly valuable in assisting the cycling and movement of nutrients around the farm, they do not replenish nutrients. Rather, exactly the same as crops, they cause the depletion of soil nutrients when they are taken off the farm and sold. So, while N can be replaced from atmospheric N (gaseous) via bacterial fixation, P, K, Mg and all the other nutrients, which only exist in non-gaseous i.e., solid or liquid, forms, can therefore only be replaced in solid or liquid forms (e.g., manure and slurry), either as biological matter or permitted mineral fertilisers. There are emphatically no other means of replenishing these nutrients at sub-geological timescales. Failure to replace these nutrients will, within a few years, inescapability lead to soil nutrient deficiency resulting in declining yields and plant and animal health, which is contrary to organic principles.

The aim for organically approved fertilisers is to allow biological soil processes (microbial activity) to progressively release the nutrients contained in the fertiliser so plants get a more balanced and continuous supply. Many of these biological processes are temperature dependent, so more plant available nutrients are released during the growing season when the soil is warmer and when plants need them, while less are released in the cold of winter when there is a greater risk of nutrient leaching and many plants are barely growing. This also means that it is not normally possible to get a 'quick response' from organic fertilisers, so if a deficit occurs it will take some time to correct. This means that it is essential to have a long-term nutrient strategy, which is also a requirement under organic certification standards. In a nutshell, a nutrient strategy is based on regular, ongoing soil nutrient analysis, coupled with nutrient budgets, which are used to determine the need to apply manures, composts and permitted fertilisers.

Soil Organic Matter

Soil fertility is linked intrinsically to soil organic matter (SOM), because it is important in maintaining good soil physical conditions (e.g. soil structure, aeration and water holding capacity), which contribute to soil fertility, and it is an important nutrient reserve. Organic matter also contains most of the soil reserve of N and large proportions of other nutrients such as P and sulphur. Typical ranges for SOM are from as little as 1.5% (of dry soil weight) in sandy soils under arable cultivation, to as much as 10% in clay soils under permanent pasture. At the upper end of this range, this can amount to between 5 and 15 t organic N/ha in the top 15 cm. Peat soils can have upward of 15% organic matter.

SOM also plays a pivotal role in soil structure management. Young SOM is especially important for soil structural development, improving ephemeral stability through fungal hyphae and extra cellular polysaccharides. To achieve better soil structure, workability and soil aggregate stability and the advantages that this conveys, frequent input of fresh organic matter is required. Practices that add organic materialare routinely a feature of organically farmed soils and the literature generally shows that, comparing like with like, organic farms have at least as good and sometimes better soil structure than conventionally managed farms.

The wider aim of soil management in organics is to create a healthy, biologically active soil flora and fauna by maintaining good levels of soil organic matter and minimising soil disturbance caused by tillage. Changing from a synthetic fertiliser regime to one based on legumes for N fixation, manures and mineral fertilisers can considerably increase soil biology, which results in many positive benefits. However, there is significant scientific evidence and farmer experience that this change takes several years and can result in an initial drop in crop yields in the first two to three years of conversion until the soil biological processes have increased sufficiently to support good yields again. This effect has also been documented when changing other production practices, for example from tillage to no-tillage. This effect is important to consider when deciding whether to continue cropping during conversion or to sow pasture. Maintaining cropping has the advantage of continuing income, but has the disadvantage of lower transitional produce prices while yields may be depressed. It is also likely that the soil will take longer to adjust to the new management system than under pasture. Growing a clover based pasture during conversion will help speed soil transformation, but unless there are stock on farm to graze the pasture, it is unlikely to generate any income so will be a financial burden in the short term. Farmer experience indicates that growing pasture during conversion has the greater longer-term advantage.

Soil Biology

The soil hosts complex interactions between vast numbers of organisms, with each functional group playing an important role in nutrient cycling: from the macro fauna (e.g. earthworms) responsible for initial incorporation and breakdown of litter through to the bacteria with specific roles in mobilising nutrients. Earthworms have many direct and indirect effects on soil fertility, both in terms of their effects on soil physical properties (e.g. porosity) and nutrient cycling through their effects on microfloral and -faunal populations (density, diversity, activity and community structure). Thus, although micro-organisms predominantly drive nutrient cycling, mesofauna, earthworms and other macro fauna play a key role in soil organic matter turnover. Factors that reduce their abundance, be it natural environmental factors (e.g. soil drying) or management factors (e.g. cultivation, biocides), will therefore also affect nutrient cycling rates. Organic farming's reliance on soil nutrient supply requires the presence of an active meso- and macro-faunal population.

The soil microbial biomass (the living part of the soil organic matter excluding plant roots and fauna larger than amoeba) performs at least three critical functions in soil and the environment: acting as a labile source of carbon (C), nitrogen (N), phosphorus (P), and sulphur (S), an immediate sink of C, N, P and S and an agent of nutrient transformation and pesticide degradation. In addition, micro-organisms form symbiotic associations with roots and act as biological agents against plant pathogens, contribute towards soil aggregation and participate in soil formation.

Generally, organic farming practices have been reported to have a positive effect on soil microbial numbers, processes and activities. Much of the cited literature has made direct comparisons between organic/biodynamic and non-organically managed soils. The evidence generally supports the view of greater microbial population size, diversity and activity, and benefits to other soil organisms too. However, little is currently known about the influence of changes in biomass size/activity/diversity on soil processes and rates of processes. Nor is it possible to conclude that all organic farming practices have beneficial effects and non-organic practices negative effects.

Soil Testing

Whatever type of fertiliser is to be applied, e.g., mineral or manure, application rates must be determined by field-by-field soil analysis coupled with nutrient budgets. It is not appropriate to apply fertilisers and manures ad hoc, as this has the potential to waste valuable and sometimes expensive fertilisers, cause leaching and run off which pollutes waterways, and cause an imbalance of soil nutrients e.g., a high K with low P. Soil tests should be viewed as a long-term strategy and investment with each field being tested every three to five years depending on intensity of production. They should be taken at the same time of year, ideally at the same stage in the rotation. These should be cross-referenced with nutrient budgets for each field, which will give a useful double check if excessive, or insufficient amounts of nutrients are being applied. If the soil tests and nutrient budgets agree, e.g., more K is being applied than removed and K soil levels are increasing then the action required is clear (do not apply any more K until the level drops and budgets balance). If they are at odds, e.g., more P is applied than removed but the P level is decreasing, this indicates a loss from the system, which requires further investigation.

Soil tests also need to be tailored to the production system. Lower intensity and livestock only farms have lower demands on their soils, and will be less affected by minor nutrient deficiencies. In comparison more intensive arable crops and very intensive horticultural crops can experience sizeable yield and quality reduction due to small deficiencies. On lower intensity farms, testing may only be required every five years with only the basic suite of tests, while three years is considered a minimum for intensive systems with the full range of tests required for horticultural holdings. Soil tests should also be viewed as an investment not a cost. While the tests may seem expensive, the information they provide can potentially increase product yield and quality dramatically which can improve profit many times that of the price of the tests. Soil tests can give some of the best return on investment of any farm expenditure.

Organic Certification Standards

With soil health being a key focus of organic systems, a significant portion of organic standards relate to soil management, particularly what materials can and cannot be used as fertilisers. Certification has three 'categories' that it puts farm inputs into, 'permitted', 'restricted' and 'prohibited'. Permitted are allowed to be used without restriction. Restricted are allowed to be used but normally only after permission has been given by the certification agent, and prohibited are completely banned. When deciding which fertilisers to use, whether it is mineral or biological, make sure you are clear which categories each is in and ensure you have the necessary permissions to use them.

Although nutrient management in organically managed soils is fundamentally different to soils managed non-organically, the underlying processes supporting soil fertility are not. The same nutrient cycling processes operate in organically farmed soils as those that are farmed non-organically although their relative importance and rates may differ. Nutrient pools in organically farmed soils are also essentially the same as in non-organically managed soils but, in the absence of regular fertiliser inputs, nutrient reserves in less-available pools might, in some circumstances be of greater significance.

Nitrogen

Nitrogen is the single most important nutrient required for herbage growth. In organic systems major emphasis is placed on N from mineralised soil organic N and from application of animal manures, but the ultimate source of nitrogen input into the system is atmospherically derived N fixed by legumes grown in pastures, the most common legume is white clover. The N resource in organic farming is therefore, a renewable resource in comparison to fossil-fuel derived artificial N fertilisers.

The primary source of N in organic farming is the fixation of elemental atmospheric N into ammonia by the bacterium Rhizobium that live in the root nodules of leguminous plants and also in the soil. Ammonia is rapidly converted into other mineral N forms, e.g., ammonium, nitrate and nitrite by other soil microbes. Nitrite is the primary form in which plants take up N although they can also take up nitrate and ammonium. The mineral forms of N are also converted into organic matter, by both microbes and plants, which is the form in which most N is 'stored' in soil. Therefore, growing legumes, in pasture, as green manures and as cash crops is essential for successful organic N management.

The amount of N which is built up by fertility building is only part of the challenge of good N management. Good management of the soil, crop and rotation is of paramount importance to maximise the efficient use of the N built up. How effectively the N is used by the subsequent crops in the rotation will depend on many factors, including: the rate of release ('mineralisation'); the efficiency of uptake by crops; the N removal in harvested products; the N return in plant residues; losses of N; timing or cropping (spring vs. winter); timing and type of cultivation and location and rainfall of the site. The rate of depletion will be reduced if manure is applied or if the rotation includes further legumes during this phase.

The factors affecting N release from the soil, interaction with crop uptake and loss processes, and the methods of predicting N release are complex. After a ley is incorporated and before the next crop can use the accumulated N, it has to be converted ('mineralised') into plant available forms (nitrate and ammonium). Some will already be in this form; most will need to be mineralised by microbial action after cultivation. Generally, the organic forms of N associated with the fertilitybuilding crop are termed 'residue N'. It should also be noted that not all of the residue N will necessarily be fixed N – some will have derived from uptake of N released from the native soil organic matter, some will be from atmospheric N deposition and some will be from soil mineral N in the soil at the time of establishment of the fertility-building crop.

The release of Nitrogen via mineralisation is performed by soil micro-organisms when they use organic N compounds as energy sources. Plant available N is a by-product of this microbial degradation.

The rate at which they undertake the mineralisation is affected by many components including; soil temperature and moisture; soil biological 'health'; Soil texture; Soil physical condition; Soil disturbance; and the type of residue.

Temporal patterns of N uptake by the crop may be particularly important in organic systems where N is released gradually by mineralisation of organic matter. For example, maximum uptake of N by winter wheat occurs in spring when soils are only beginning to warm and mineralisation is still slow. This is likely to limit the supply of N at a critical time for wheat crops on organic farms.

Mycorrhizal fungi have been shown to absorb and translocate some N to the host plant, so maintaining good mycorrhizal fungi populations can be beneficial in N utilisation.

A balance therefore has to be struck between exploitative and restorative phases of the rotation to ensure that as N drops during the exploitative cropping phase it is replaced by restorative crops and pasture so maintaining N levels over the rotation as a whole.

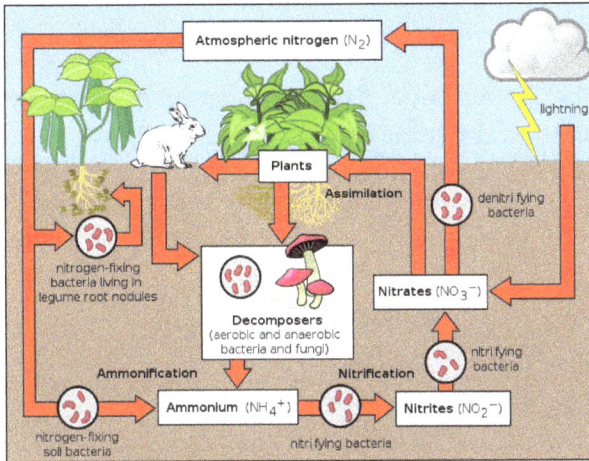

Simplified Diagram of the Nitrogen Cycle.

Predicting the actual amount of nitrogen fixed is notoriously difficult as it depends on many factors including legume species and cultivar, proportion of legume in the ley, management, weather conditions and the age of the ley.

Nutrient supply to crops depends on the use of legumes to add nitrogen to the system and limited inputs of supplementary nutrients, added in acceptable forms. Manures and crop residues are carefully managed to recycle nutrients around the farm. Management of soil organic matter, primarily through the use of short-term leys, helps ensure good soil structure and biological activity, important for nutrient supply, health and productivity of both crops and livestock. Carefully planned diverse rotations help reduce the incidence of pests and diseases and allow for cultural methods of weed control.

White Clover

White clover, has the unique ability to fix atmosphere nitrogen, and can transfer the equivalent of 80-100 kg N/ha to the growing grass. In addition to contributing to herbage growth, a proportion of the atmospherically fixed N is stored in the roots and stubble or is immobilised in the soil organic matter i.e. it contributes to the onger term build-up of soil organic matter and soil nitrogen status. It is therefore vital that if organic grass farming is to be productive, white clover, and /or red clover must be introduced into pastures.

It is essential to maintain clover in the pasture so management practises must be tailored to achieve this objective – poaching must be avoided at all costs. On wetter farms the persistence of clover is problematic and low cost methods of clover introduction must be implemented.

There are losses of nitrogen both as gaseous emissions to the atmosphere and leaching during spreading and storage of animal manures. It is essential to minimise these losses by covering the manure during storage to minimise the effect of leaching by rainwater. Losses to the atmosphere can be reduced by using a band-spreader or shallow injector rather than a splash plate vacuum tanker.

Table: Examples of the amount of N fixed per hectare per year by mixed pasture and two crops and the amount remaining after harvest.

Crop	Amount fixed kg N/ha/yr	Amount remaining after harvest kg N/ha/yr
Red clover silage	160 to 450	50 to 150
White clover & grass silage	70 to 420	20 to 180
White clover & grass grazed	60 to 250	50 to 210
Forage peas	80 to 290	40 to 110
Field bean grain crop	200 to 380	90 to 150

Phosphorus

Phosphorus is an essential element for plant and animal life. In animals, P is required for bone formation and a deficiency can cause osteomalacea. 'Pica' or depraved appetite has been noted in cattle when there is a deficiency of P in the diet. Low dietary P may also be associated with poor fertility, and apparent dysfunction of the ovaries causing inhibition, depression or irregularity of oestrus. Phosphorus is a key nutrient for the Rhizobium bacteria that fix atmospheric N into soil mineral N, so good P levels are critical for maximising N fixation.

The quantities and forms of P in soils depend on the degree of weathering, the nature of the soils parent material and their management. Phosphorous exists in both organic and inorganic forms in soils, the inorganic forms being derived primarily from the weathering of soil minerals. The pathways and mechanisms through which P evolves into its different forms are both complex and dynamic.

The main sources of P on organic farms are on-farm produced manures, slurry and other biological materials. It is therefore important to utilize this resource where it exists and certification standards stipulate than the primary source of P should be from on-farm manures. However, the P content of manures varies considerably, for example 0.1% for watery slurries to 1.4% in turkey litter. Brought in feed and straw also contains P, for example, wheat straw can contain 0.15%, barley 0.25% and oat 0.22% P, and their grains contains around 0.4% P. The nutrient content of such materials should be included in nutrient budgets.

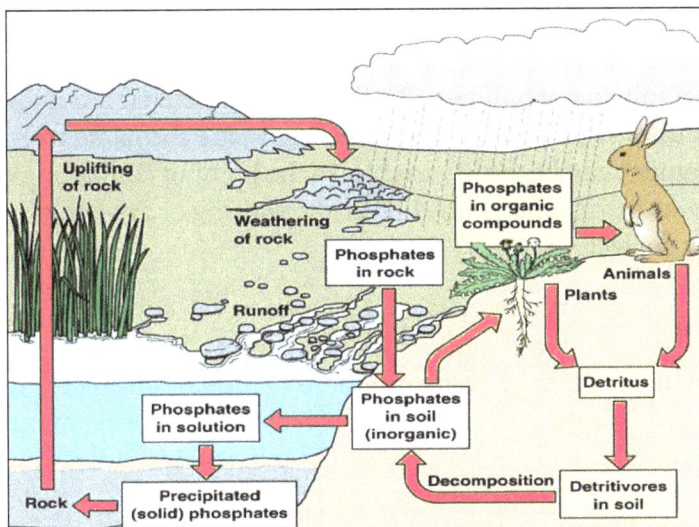

A Simplified Farm Phosphorous Scheme.

The main mineral source of P available in Ireland is natural rock phosphate called ground rock phosphate (GRP) that has a P content of around 13 to 40% depending on source. This should be generally available as it is used by the forestry industry; however, it may not be stocked by farm fertiliser merchants and may have to be ordered in. GRP is not suitable for soils above 7.5 pH as they are insufficiently acid to decompose GRP and release the phosphate. Organic standards recommend calcined aluminium phosphate rock for soils above pH 7.5; however, this is not currently available in Ireland and would have to be ordered from the United Kingdom. In this situation, specialist advice should be sought. However, clover is known to acidify the area around its roots which assist it absorb soil P, which means that GRP may decompose above pH 7.5 under clover rich pasture. Unfortunately, there is little research studying this effect, so no advice regarding it can be given.

GRP contains about 35% calcium (Ca) which helps balance the acidifying effect of the P compounds. This means that most GRP's have little effect on soil pH. This contrasts with super-phosphates, which have an acidifying effect. GRP also contains useful amounts of some trace nutrients.

GRP is a slow release fertiliser as the P and calcium are released by soil acids and biological activity. About a third of the P is released each year, which means that it takes about three years to decompose fully. This also means that it takes at least six months to a year for the fertilisation effect to be seen. This is in contrast with super phosphates, which are water soluble, rapidly plant available and give a quick crop response. This delayed fertilisation effect must therefore be planned for, with the GRP applied a year before the P is required. The slow release action can also help reduce P loss to the environment.

Phosphate rocks naturally contain heavy metals, particularly cadmium, in varying amounts. Certification standards require that lower cadmium content rock should be used and depending on quantities used, soil analysis may be required to ensure that cadmium levels in soil do not build up. Basic slag is a restricted input for P, K and micro nutrient sources in organic agriculture. However, some forms/sources are prohibited and the P level content can be quite variable, to the point of containing only traces of P. Always consult your certification body before purchasing or applying basic slag and ask for a full nutrient analysis of the product.

The sources of P loss and the pathways of P loss are the subject of considerable controversy. There are three main sources of P losses: 1) Seepage of soiled water from farmyards is one of the main culprits and there is little doubt that if farmyard design and maintenance were improved, there would be less P pollution of our waterways. 2) Slurry spreading, in itself will not lead to run-off of P. It is spreading slurry at the wrong rates, or in the wrong place, or at the wrong times that lead to slurry finding its way into drains, rivers and streams. Spreading slurry at reasonable rates, during the grass growing season in places where there is no risk of runoff into rivers/lakes will ensure no loss of P. 3) Elevated soil P levels. There is evidence that increasing soil P levels can lead to increased P runoff in areas where run-off to water-bodies is possible.

Potassium (K)

The behaviour of potassium K in the soil is in many ways similar to that of phosphorous. Inputs of K from the atmosphere are usually negligible; therefore its quantities and forms in the soil depend on the nature of the parent material, the degree of weathering, and their management. It exists in the soil partly in the structure of soil minerals (lattice K), partly fixed inside clay minerals (fixed K), partly on cation exchange sites (exchangeable K) and partially as K+ ions in the soil solution (soluble K).

Soil K levels will drop dramatically from grounds where silage/hay is continuously cut and manures are not reapplied to.

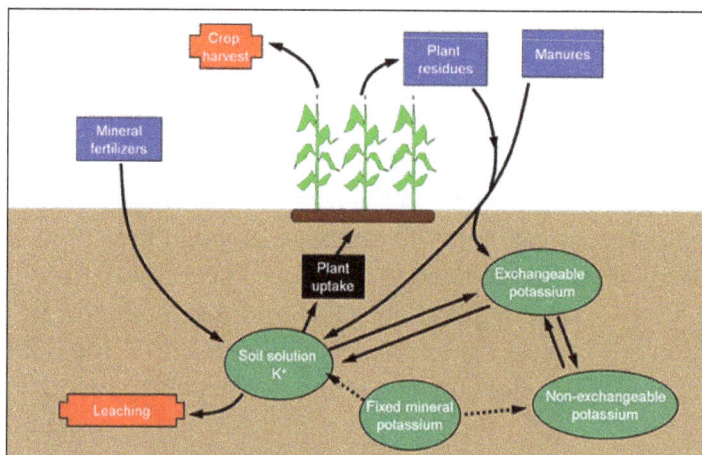

Simplified Potassium Cycle.

As for P, the main source of K is on-farm produced manures, slurries and compost, which have K values ranging from 0.2% for cattle manure to 2% for poultry manure. Brought in materials, such as manure and compost as well as production inputs such as feed and straw can contain useful amounts of K, for example, grain contains 0.4 to 0.6%, with straw containing 8 to 15% K depending on species and K status of the soil they were grown on. As for all other nutrients brought in materials should not form the basis of an organic farm's nutrient sources, rather effective internal cycling should be the main approach, supplemented with brought in materials to balance those lost from the system. For K this is principally in crops and livestock as K does not readily leach from soils. As for P, detailed records of all K sources and losses must be recorded to allow accurate nutrient budgets to be compiled.

Alternative sources to biological materials include wood ash, which has around 1 to 7% potash, and is permitted under certification rules. However, it must be mixed with compost or manures, the wood must be untreated and sources are limited. The increased interest in wood chip and other biomass boilers may see this situation change.

If K soil levels are lower than the desired value and on-farm sources are insufficient to meet needs, an application to your certification body to use Potassium Sulphate (K_2SO_4), more commonly referred to as sulphate of potash (SOP), must be made. However, ongoing use of SOP may be restricted depending on circumstances. There are a limited number of SOP supplies in Ireland; these are outlined in the table below.

Sources of Sulphate of Potash
Sulphate of potash K_2SO_4 (SOP) (a generic and widely available product), 41% K (as 50% K_2O/potassium oxide), 18% S (as 45% SO_3/sulphur trioxide)
Sulphate of potash with magnesium and sulphur: (e.g., Patentkali), 25% K (as 30% K_2O/potassium oxide), 6% Mg (as 10% MgO/magnesium oxide), 17% S (as 42.5% SO_3/sulphur trioxide)
Sulphate of potash with sodium and sulphur (e.g., Magnesia-Kainit), 9% K (as 11% K2O/potassium oxide), 3% Mg (as 5% MgO/magnesium oxide), 20% Na (as 29% Na_2O/sodium oxide), 4% S (as 10% SO_3/sulphur trioxide)

The two propriety products, Patentkali and Magnesia-Kainit, have EU level organic approval at the time of writing while SOP may require case-by-case organic certification body approval depending on source. It is essential to check the current approval status before purchasing and/or using such materials.

All three forms are highly water-soluble so they will be rapidly taken up by any plants present before the K is absorbed into the soil. Potassium leaching is less of a problem than P and N, except if applied under very unsuitable conditions, e.g., where there is active surface run off.

Lime Requirement and pH

Our high rainfall sometimes causes an increase in the acidity of our soils, regardless of the system of farming. Maintaining soil pH between 6.3 and 6.6 is essential for many microbiological processes in the soil and is therefore essential for nutrient cycling. It is also absolutely essential to maintain pH at optimum levels to allow the satisfactory establishment and growth of clover.

The maintenance of optimum pH is as, if not more, important in organic production than non-organic. The correct pH is essential for optimal soil biological activity, especially the mineralisation (decomposition) of organic material into plant available nutrients, which is the main source of N in organically managed soils. High pH also limits the rate of rock phosphate dissolution, so pH in organic system should tend towards the acidic rather than alkaline, within the range of 6.2 to 7.0. High organic matter levels also helps dissolve rock phosphate due to the presence of biological acids.

Most sources of lime are permitted in organic systems, but it is important that the particular product to be used is either organically certified or that your certifier confirms that the source is permitted. Similar to rock phosphate, lime is relatively insoluble and the calcium and other constituents have to be released by biological decomposition, which means it is released slowly, often over five years or more. The finer the lime is ground the quicker it will decompose and the quicker it will raise pH. However, finer material is more expensive and difficult to apply (it blows away), so that in most cases it will be more economic to apply larger amounts of courser material to rapidly rectify high pH than smaller amounts of finer material.

Determining when to apply lime to maintain an optimal pH is achieved by the use of regular soil testing, just as for soil nutrients. With lime taking several years to have its full effect on pH it needs to be used 'pre-emptively' before pH drops too low.

Manure Management

The components of a manure recycling system on grassland farms are manure production, storage, application and utilisation. The level of control that a grassland farmer exerts on the production of manure is limited. However, farm management practises that optimise the quantities of bedding materials used will significantly reduce the quantities of manure production. For example, significantly less straw is used in cubicle systems compared with loose housing (120 versus 530 kg) over a 180 day winter. For liquid systems, reducing the amount of water entering the slurry store will reduce the volumes produced.

Animal manures, composts and brought in biological materials must be handled carefully to minimise nutrient losses. The primary cause of nutrient loss, mainly N and K, is due to rain exposure. Biological material must therefore be protected from the rain, for example, by being in a shed, under plastic sheeting or other suitable rainproof covering. However, tight fitting, air proof covers, such as plastic sheets, can impede airflow into and out of material that is undergoing rapid aerobic 'hot' composting. These are mostly best left uncovered until the composting process is finished unless there is so much rain that the compost cannot absorb it. Storing biological material in fields and/or on soil is not recommended due to the potential for leaching and environmental pollution. Ideally, material should be stored on an impermeable base that allows the collection of any liquids produced. Silage pits and stock sheds are well suited for this; however, silage pits must be thoroughly cleaned before being used for silage.

On-farm produced livestock manure and slurry can be valuable sources of N, especially for early season pasture growth. This is because white clovers need temperatures above 20 °C to grow vigorously and fix significant amounts of N, so Irish spring soil temperatures are too mostly to cold for white clovers to meet grasses' high N demand in spring. Slurry and manures can therefore help meet this spring N requirement shortfall. Organic standards state that the total amount of manure applied to the holding must not exceed the equivalent of 170 kg of nitrogen per hectare of agricultural area used, calculated over the whole area of the holding or linked units.

Nutrient Content of Animal Manures

The quantity of nutrients in manures varies with type of animal, feed composition, quality and quantity of bedding material, length of storage and storage conditions. Surveys on manure composition suggest that the dry matter and nutrient contents of manure can vary up to ten fold even from the same type of animal. In organic systems it is particularly important to conserve manure nutrients for both economic and environmental reasons. On mixed and livestock farms animal manures an important means of re-distributing nutrients as it is important to ensure that excessive fertility is not built in some fields at the expense of others. Manure use should be planned with regard to both farm system and field nutrient budgets. Manures play a key role in fertility building and maintenance in many organic rotations. Understanding their nutrient composition and nutrient availability is therefore important for optimising their use on farm. Typical estimates of total N, P & K contents of common animal manure types are outlined in the table below.

Typical N, P & K levels in animal manures			
Animal Manure type	N kg/t3	P kg/t3	K kg/t3
Farmyard Manure	4.5	1.2	6.0
Cattle Slurry	5.0	0.8	4.3
Compost	7.5-15.0	1.0 – 2.0	7.0
Poultry Broilers Deep Litter	11.0	6.0	12.0
Layers	23.0	5.5	12.0
Turkeys	28.0	13.8	12.0

Manures are a valuable source of nutrients (and organic matter), and can be seen as a method of transferring nutrients around the farm (for home produced manures) or as a method of importing fertility (imported manures or composts). Good manure management offers a 'win-win' opportunity: benefits to soil fertility and benefits to the environment (less pollution).

When not applied appropriately, animal manures applied to agricultural soils can be significant contributors to nitrate leaching. Large amounts of N can also be lost from the soil in surface run-off when heavy rain falls in the first few days after slurry application. It is the 'readily available' nitrogen fraction that is most at risk from leaching: ammonium-N, uric acid-N (poultry manures) and nitrate-N (generally only trace amounts in most manure). In organic farming most manures are produced from either slurry or straw-based systems.

The straw based systems have a relatively small readily available N content, thus presenting a small nitrate leaching risk. Some manures are also composted, which tends to reduce their ammonium N content still further. However, it should be noted that nitrate can accumulate during composting and it may be that well-composted manures have potential to leach substantial nitrate (either from an uncovered heap or after application to land in autumn). Another route for N loss is that of direct run-off of N in leachate from manure stores. Clearly, manures have to be managed in such a way as to minimise this risk by having facilities to collect the leachate. Covering the manure will not necessarily eradicate the risk, because much of the N is contained in the liquid that leaks from the FYM heap in the first few days). The N content in leachate leaving the heap declines with time, because the readily available N becomes assimilated into the organic fraction of the manure heap.

Ammonium N, the principle plant available source in animal manures is also susceptible to losses. The ammonium-N is readily converted to ammonia gas (NH_3), which can be lost to the atmosphere. Exposure of slurry or farmyard manures to warm, windy and sunny conditions at time of application promotes high ammonia losses, therefore applications in a manner that minimises N loss will maximise the fraction available for crop growth.

Timing and Methods of Application

The application rate of manures should reflect the nutrient concentration of the manure and crop requirements. Slurries and manures should be analysed to provide some indication of nutrient content.

1. Apply at a time when crop demand is high in spring. Where soil conditions allow, aim to have 70% applied by the end of April. Opportunity for spring application on heavy soils may be increased by using application methods that reduce soil compaction. Early application provides nitrogen for grass growth before clover starts to fixate atmospheric nitrogen from mid May onwards.

2. Application in dull, overcast or misty conditions will result in lower ammonia losses compared to application in warm, dry or sunny weather.

3. Animal manures should not be applied when the soil is:

    •   Waterlogged,

    •   Flooded,

    •   Frozen or covered in snow,

    •   Heavy rain is forecast within 48 hours,

    •   The ground slopes steeply giving a significant risk of water pollution.

Slurries with low dry matter contents will percolate and wash into the soil more quickly than material with higher dry matter, therefore reducing the duration of exposure to air. Dilution of slurries with soiled water will increase the volume to be managed and therefore increase spreading costs; however a 1% reduction in slurry dry matter percentage will decrease ammonia losses by approximately 5 to 8%. Using application methods such as band spreading, trailing shoe and shallow injection place the slurry in bands or lines rather than on the entire surface as with the conventional splash plate method.

Composting: On Organic Farms

Composting is recommended in organic farming as a management tool for controlling weeds, pests and diseases. Organic standards promote composting, anaerobic digestion, aeration of slurry and correct storage of manure. These treatments greatly reduce pathogen loads in manure by increasing the range of biological activity, which helps to suppress pathogenic microbial populations, and by heat pasteurisation. A well-managed aerobic digester or aerobic compost heap will reach temperatures of 55°C to 65°C, and will be maintained at this temperature for three days to destroy weed seeds and pathogenic bacteria. In addition, aerobic composting results in the stabilisation of nutrients, giving the compost nutrient release characteristics that are more in tune with the demand of crops throughout the seasons.

True composting of manures, i.e. aerobic decomposition at temperatures of around 60 Deg C, results in fundamental physical and chemical changes to the manure. Composting results in some losses of nitrogen through volatilisation in the form of ammonia however the soluble nutrients, particularly nitrogen, are stabilised and hence subsequently less liable to leaching. Composted manure thus has a more longterm role in building soil fertility, and has been shown to be more effective in building soil microbial biomass and increasing activity than un-composted manure.

Importing Animal Manures

Many horticultural holding import manure from other organic farmers and conventional farmers. However, the opportunity to import from other organic farmers is limited due to the requirement for the manures to replenish nutrient levels on these farms.

Manure can be imported from conventional farms provided that it complies with the following conditions:

- Not from animals reared intensively i.e. above 170 kg/ha of organic N,

- Not from livestock fed diet which includes a GM component,

- Sewage sludge,

- Composted for three months when it arrives on the organic unit,

- The total organic N for the unit does not exceed 170 kgN/ha.

Trace Eliments on Organic Farm

On well established organic farms, sound agricultural procedures should render normal supplementation unnecessary. The aim should be to reduce and eventually eliminate imported minerals by growing appropriate varieties of grasses and herbs necessary. Routine additions to food of mineral and vitamin supplements are prohibited.

Animal Requirements

The minerals and trace elements that are essential for animal health are calcium (Ca), phosphors (P) potassium (K), magnesium (Mg), nitrogen (N) sodium (Na), and sulphur (S).

The trace elements that are essential for animal health are iron (Ra) manganese (Mn), copper (Cu), zinc (Zn), Cobalt (Co), selenium (Se), iodine (I), and molybdenum (Mo). The trace element baron is not essential for animal health, but its absence can affect the survival of certain pasture species, in particular clover.

In organic farming the main sources of minerals and trace elements for livestock is the soil itself and the recycled materials generated from the farm. Outlines the concentrations of trace elements normally found in soils and herbage in organic farming. Herbage values are given as total concentrations. While those for soil are given as plant available levels. Soil levels for Fe, Se and I are not given soil values for Fe bear little relationship to herbage values. Se and I are easily leached and the soil concentrations are reduced by rainfall. Iodine values are also affected by proceeding to the sea. The factors that influence the availability of trace elements are numerous, interpreted and interrelated. The factors are soil type, drainage, grazing, management, pH and organic matter, grass species.

Soil Types and Drainage

Free draining sandy or gravely soils are low in trace elements. The low status of these soils is associated with low clay, oxide and organic matter control rather than with the leaching of trace elements from the soil profile.

Poorly drained soils affect the trace element supply to livestock in two ways. Firstly, poor drainage can increase the concentrations of some trace elements soil and in pasture species. Secondly, poor drainage can increase soil ingestion by livestock and this can lead to further deficiency issues.

Table: Range of trace elements in the soil and herbage of organic grassland.

Trace Elements	Soil Concentration (mg/kg)	Herbage Concentrations (mg/kg)
Iron (Fe)		50-150
Manganese (Mn)	50-250	25-250
Copper (Cu)	5-8	7-9
Zinc (Zn)	2-10	20-60
Cobalt (Co)	3-10	0.05-0.15
Selenium (Se)		0.03-0.30
Iodine (I)		0.08-0.30
Molybdenum	1-2	2.0-5.0
Boron (B)	0.8-1.5	2.0-10.0

Lime Application

The soil pH influences the availability of trace elements for uptake by pasture species. A soil pH of 6.5 is considered to be the optimum for soils containing trace elements in well balanced amounts. All soil pH values below 6.5 the availability of Mo and Se is reduced and the availability of Fe, Mn, Co, Zn and B is increased. Low soil Mo can lead to poor clover establishment, while Se deficiency can lead to cardiovascular problems in calves and afterbirth retention in cows. Soil pH above 6.5 the availability of Fe, Mn, Co Zn and B is reduced and that of Mo and Se is increased. In Irish conditions, leaching by rainfall can cause the pH to fall. This must be rectified by lime application.

Soil Organic Matter

Organic matter (humus) is an important constituent of every soil because it is the most chemically and biologically active of all the soil phases. It is of particular importance in organic farming because the farmer is dependent on inherent soil properties to support crop and livestock growth. The beneficial effects of organic matter in soil are as follows:

- Improves soil structure.

- Improves the water holding capacity of a soil.

- Absorbs from 2 to 30 times as many cations as other active soil phases.

- Depending on content accounts for 30 – 90% of the adsorption power of mineral soils.

- Improves the ease of interchange of plant nutrients with soils.

- Can extract plant nutrients from the mineral phases of the soil.

The organic matter content of a soil is dependent on climate and soil type. In wet temperate regions soil organic matter is high. In arid regions soil organic matter is low. In heavy clay soils the organic matter tends to be high. In light, sandy or gravely soils organic matter is usually low. The range of values across regions and soil types varies from less than 1% to over 15%. Values above 15% indicate poor drainage. At soil organic matter values greater than 20% soils are classified as peat soils. Intensive cultivation and liming causes a reduction in soil organic matter. Soil organic matter content is greater under permanent grassland than in arable land. In Ireland, in most grassland soils the organic matter is highly satisfactory at 8-10%.

Soil organic matter is strongly associated with the transfer of trace elements from the soil to the plant. Although trace elements may be fixed by the well decomposed organic materials such as the humic acids, the less well decomposed organic acids, polyphenois, amino acids, peptides, proteins and polysaccharides carry trace elements in loosely bound chelated form which are easily available to plants. From 98 – 99% of Cu, 84-99% of Mn and 75% of Zn are carried on organic complexes within the soil. It is clear from the above that practices that build up the organic matter in a soil will also increase the mineral and trace element supply to livestock. A low livestock density (1.0-1.5 LSU/ha) prevents over grazing and a reduction of organic matter input to the soil. The soil pH should be at 5.5 or slightly below.

Because of its excellent growth characteristics perennial ryegrass is likely to remain the mainstay of productive organic farming. However its mineral composition is low compared to other grass species. Among the nine elements listed perennial ryegrass has the highest concentration of I. Rough stalked meadow grass contains highest concentrations of Fe, Mn, Cu, Co and Mo. Red and white clover has higher concentrations of Fe, Cu, Zn, Co and Mo than perennial ryegrass.

Trace elements concentrations in pasture species (mg/kg)								
Species	Fe	Mn	Cu	N	Co	Se	I	Mo
Perennial Ryegrass	55	87	5.0	23	0.05	0.03	1.46	0.31
Rough stalked meadow grass	153	119	12.3	16	0.06		0.14	0.59
Italian ryegrass	148	65	4.3	11	0.05		0.85	0.29
Timothy	34	97	5.5	24	0.04			0.35

| Red Clover | 74 | 74 | 11.9 | 31 | 0.11 | 0.01 | 0.40 | 0.35 |
| White clover | 88 | 87 | 8.4 | 28 | 0.10 | 0.10 | 0.91 | 0.41 |

It can be concluded that to achieve the best balance of trace elements in pasture it is advisable to have perennial ryegrasses and clovers, and contributions from other species, if necessary.

Lack of Trace Elements in Soil

Where there is a known dietary deficiency in home grown feeds, or as a result of soil deficiencies, on there is veterinary evidence (e.g. blood, soil, or herbage) of a deficiency within the livestock, restricted supplementation will be permitted with approval from the Organic Certification Bodies. Only minerals and vitamins from permitted sources may be fed. All supplements must be GMO free.

Role of Rotation

Rotations are a key technique for managing overall nutrient supply on organic farms, particularly for mixed and stockless systems. The clover-grass ley is the most important fertility-building phase, principally for nitrogen, but it is also an appropriate time to be applying P and potentially lime. Applying P to the start of the pasture phase means that it will help maximise N fixation. At the same time, the higher levels of soil biological activity found under pasture plus the presence of clover will allow for the most effective breakdown of the mineral P and conversion into biological forms that will feed the following crops.

Likewise, applying lime at the start of the pasture phase allows it to be incorporated into the soil by tillage and it allows time for it to be converted from mineral to biological forms while optimising soil pH under the ley, thereby maximising pasture productivity. For lime sensitive but nutrient demanding crops such as potatoes putting lime on at pasture inception means they can be planted soon after pasture to make the most of the higher nutrient levels while minimising the risk of lime damage. Putting lime on at pasture termination risks over-stimulating organic matter mineralisation in addition to that caused by tillage, which could result in substantial nitrogen losses.

As the approved mineral forms of K are highly water soluble, putting them on at the start of the ley will result in them being converted to biological forms or entering the less available K soil reserves. If K levels going into pasture are below optimum, then K should be applied to optimise pasture production. If K levels are within recommended levels, consideration should be given to reserving K application for crops with a higher K demand, e.g., root crops.

The cropping sequence following a ley starts with the most N and other nutrient demanding crops, typically wheat, maize and root crops. These are followed by less demanding crops, such as brassicas, and ending with crops that are efficient nutrient scavengers, e.g., oats, or N fixing legume crops. The addition of manures (FYM, slurry, compost) is generally reserved for the last half of the rotation. While this can be an effective way to 'top up' soil nutrients it is unlikely to be able to build soil N up to the levels found after the ley.

Unless a straightforward rotation is being followed, crops should be matched to fields at the start of each year, based on each field's history, crop needs and market requirements. This has to include an evaluation of the field's nutrient and general soil conditions, plus presence of pests, disease and weeds. This is then matched to the crops nutrient and soil condition requirements, the length of rotational gap it needs in relation to pest and disease carry-over and how competitive it is with weeds. Rotations are therefore not pre-ordained and fixed, but highly flexible and changed on a yearly basis.

Where insufficient soil nutrient replacement has taken place, and soil nutrient or pH levels have fallen too low, speedy action to replace the depleted nutrients must be taken. Probably the best option in such situations is to establish a grass clover pasture rather than attempting to crop depleted land. This is because yields and the nutrient content of crops are likely to be poor; the latter is clearly contrary to organic standards and the former means that profit will be reduced. It is also likely to be harder to rectify nutrient deficiencies under crops that are predominantly removing nutrients compared to pasture where removal will be lower and livestock assist with nutrient cycling through dung and urine. Other issues that make pasture the preferred crop when rectifying nutrient deficiencies are nitrogen fixation by clover, soil structure improvement by grass roots plus the absence of tillage, all of which will help build organic matter and biological activity resulting in most cases in faster improvement in soil health than under cropping.

However, if soil nutrient deficiency is severe, great care must be taken to ensure stock do not also develop nutrient deficiency, e.g., by providing supplementary fodder and/or mineral licks, the latter must however be approved by your organic certification body.

Soil Fertility in Organic Farming Systems

In organic farming systems, soil fertility means more than just providing plants with macro- and micronutrients. Effective fertility management considers plants, soil organic matter (SOM), and soil biology. Ideally, organic farming systems are designed to enhance soil fertility to achieve multiple goals. Important goals include: the protection and, if possible, improvement of soil physical condition so that the soil supports healthy plants and soil-dwelling organisms and has the ability to resist and recover from stresses like flooding or aggressive tillage; the maintenance of soil buffering capacity to minimize environmental degradation caused by soil loss or soils' failure to filter nutrients or degrade harmful compounds; and increased water and nutrient use efficiency by increasing biological fixation and retention of needed nutrients while reducing their loss from the system to the extent possible. Organic farming systems are designed with the aim of maintaining nutrients in organic reservoirs or in bioavailable mineral forms instead of just supplying nutrients through frequent fertilizer additions. This is achieved by cycling nutrients through organic reservoirs. Soil fertility is improved by organic matter management and not through input substitution.

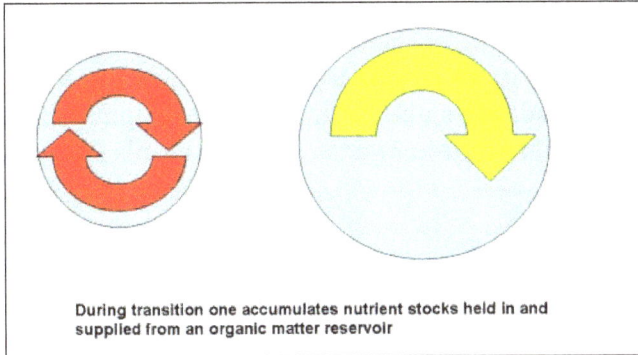

During transition one accumulates nutrient stocks held in and supplied from an organic matter reservoir

This Figure depicts idealized changes that occur where fast mineral nutrient cycling (depicted by red arrows) in depleted organic reservoirs (depicted by blue sphere) is altered by improved management to result in slower cycling rates (yellow arrow) within an enlarged reservoir.

The intention of organic systems is to manage the full range of soil organic and inorganic nutrient reservoirs and prevent unwanted loss by retaining them in forms that can be accessed by crops through biological leveraging. One objective during transition is to enhance symbiotic associations between plants and their microbial partners, organic reserves, and the physical environment.

This holistic view is the basis for the soil fertility management practices used in organic agriculture. There are four soil fertility management practices typically used in organic cropping systems that determine the cycling and availability of nutrients in the soil:

1.   Use of organic residues as soil amendments or sparingly soluble minerals;

2.   Use biological N-fixation as the major N source;

3.   Use of a rotation that includes active plant growth (cover crops, intercrops etc.) as much as possible and that minimizes bare fallow;

4.   Plant species are diversified in space and time to fulfill a variety of functions (minimize weeds and pests, support below-ground processes, erosion control, N fixation, build SOM etc.).

Organic Fertility is not a Matter of Input Substitution

Organic farming systems cannot rely on use of soluble, inorganic nutrient sources. Conventional fertilizer management guidelines hinge upon assessments of plant-available N and P combined with empirical fertilizer addition studies that are able to provide estimates of the amount of fertilizer required to achieve yield goals. Although many organic producers do use soil testing to assess soil nutrient levels, they report that while these tests often indicate that plant-available N or P may be limiting, their yields do not reflect these soil test results. There are several reasons to expect that organic production systems require their own suite of management tools. First, organic soil amendments vary in quantity and quality. Second, the condition of the soil resource plays a large role because it supplies and recycles added nutrients. Results can vary greatly after organic materials are added for a variety of reasons, some of which, can be managed. Current research on organic fertility management is looking closely into avenues for soil testing and management of amendments and soil biology to optimize fertility by taking into account the synergisms that occur in biologically active soils.

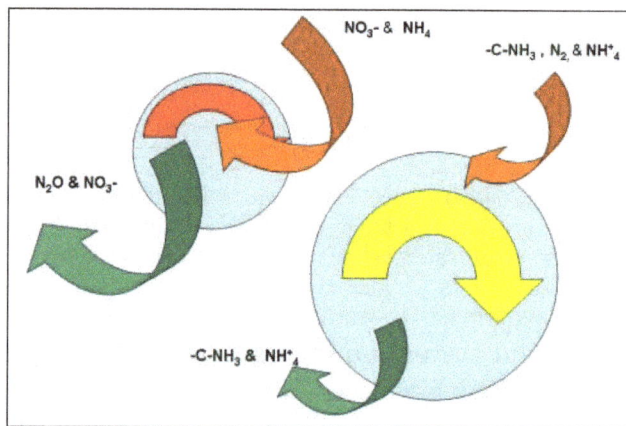

The figures above depict changes in N cycling in systems where N is supplied in inorganic fertilizers to systems that rely on organic sources. The size of the blue sphere represents the active or labile fraction of soil organic matter and the thickness and coloration of the arrows reflect the size and cycling rates of nutrient reserves. Red arrows identify pools that cycle more rapidly than yellow.

Avoiding Problems of Nutrient Over-addition with Biology

Loss of nutrients applied to agricultural soils causes environmental harm. Off-site problems caused by over-application of nutrients are better recognized than are problems caused on-site. Conventional agriculture is cited as the primary source of non-point source and P pollution that contributes to myriad environmental and health risks. Problems of over-application in organic systems vary; probably P over-additions are most widespread where manure is readily accessible. This is because the ratio of P to N in manure exceeds that required by the plant.

Over-addition of N, particularly in readily available forms, is a common problem in agriculture. Over-addition in organic systems can occur in situations where leaching is restricted (e.g. in greenhouses) or after N rich cover crops or manures are applied. The notion that N surplus promotes microbial activity that contributes to nitrate leaching and nitrous oxide emissions, two important environmental problems, which works against organic matter storage, is finally being recognized by main stream scientists.

Excess nutrients can also increase plant susceptibility to pathogens and arthropod pests and can also lead to increased weed competition. Tendencies toward nutrient leaching and the ability to hold and retain nutrients vary with soil types and climate. Soil texture and CEC are related to this, with nutrient storage capacity increasing with soil clay and silt contents and cation exchange capacities.

These problems and those associated with herbicide and insecticide application can be avoided by managing soils biological activity to enhance:

1. Reliance on N-fixation as a source of N,

2. Mycorrhizal associations,

3. Plant-induced liberation of nutrients,

4. General suppression of soil borne disease, and

5. Decay of weed seed and inhibition of weed seed germination.

There is growing acceptance that maintenance of high levels of available nutrients works against these processes and accelerates organic matter decay. In addition, nutrient excess or imbalance can compromise plant and animal health. Excess tillage can have a similarly undesirable effect on soils and soil organisms. Of course tillage is an important tool for fertility management. Perennial sods are mechanically killed, cover crops are plowed in, and manures are incorporated. The figure below shows how tillage has been argued to alter the soil food web. Heavy reliance on tillage in organic farming systems can only be maintained without harm in systems that include adequate plant cover and where tillage is timed to avoid compaction and erosion. Soil texture, slope, and climate all influence the degree to which tillage can or can not be safely tolerated.

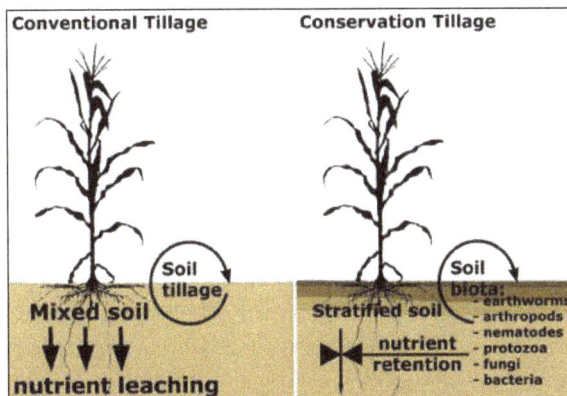

The increased soil stratification and size and activity of soil organism populations under conservation tillage compared to conventional tillage lead to increased nutrient retention.

This is why organic strategies for fertility management focus as much or more on crop rotation and tillage practices than they do on nutrient dense soil amendments. The contributions of carbon-rich amendments and roots to soil fertility are recognized within organic systems.

Soil Conservation Practices in Organic Crop Production

Conserving Soil in Organic Crops

Soil is the production base of all agricultural systems. Soil conservation is one of the pillars of sustainability. Soil quality has deteriorated due to wind and water erosion, as well as farming practices that reduce organic matter content. Soil conservation promotes practices that stop the decline and, over time, improve soil quality. In general maintaining soil cover, living plant roots and limiting soil disturbance are the best ways to limit erosion.

Conservation practices generally reduce wind erosion, reduce rate and amount of lateral water movement in fields, and increase soil organic matter levels. There is not one way of conserving soil that is well suited to all situations, due to differences in soil type, topography, type of farming operation and climate.

In organic crop production, the challenge increases, because conservation practices that involve pesticides are not an option. In every case, it is generally more effective to use a number of conservation strategies. For example, rather than plant a large number of shelterbelts on a particular field, it may be better to plant a lesser number and reduce tillage operations or include green manure more often to keep soil surface covered. The challenge is determining which conservation practices are best adapted to each farm.

Crop Residues

Crop residues include roots, chaff, stems and leaves are left after a crop is harvested. Crop residues are the prime source of organic matter replenishment. These residues improve several soil properties, such as water infiltration, water storage and particle aggregation. Crop residues also contain nutrients, including nitrogen, phosphorus, potassium, sulphur and micronutrients. Stubble left standing overwinter will trap snow and slow evaporation of soil moisture in spring, as well as prevent erosion. Moisture conservation is important to soil conservation because the additional moisture will improve crop growth the following year.

The amount of residue produced and the rate of decay vary among crops. The combination of these two factors determines the quality of residue in relation to its value for soil conservation. Field pea and flax generally produce about half the residue of cereals and canola. Residue from canola, field pea, lentil, and sunflower usually decays very rapidly (particularly when tilled), leaving the soil surface unprotected.

Conventional tillage summerfallow after oilseed or pulse crops is not recommended, unless an extensive system of wind barriers will prevent wind erosion and water erosion is not a concern. Even where recropping is practiced, excessive tillage after oilseed and pulse crops can promote serious erosion.

Cereals usually produce an acceptable level of crop residue that decays at a moderate rate. When designing a crop rotation, factors of this nature should be considered.

Table: Approximate surface cereal residue levels required for erosion control.

Wind erosion control	
Soil texture	Cereal residue required (lb./ac.)
Medium (loam)	900
Fine (clay)	1350
Coarse (sandy)	1800
Water erosion control	
Field slope	Cereal residue required (lb./ac.)
Gentle (6-9%)	700-1000
Moderate (10-15%)	1,000-1,500
Steep (16-30%)	continuous grass
Very steep (>30%)	native vegetation

Tillage

Crop residue conservation during tillage is affected by equipment type, speed, depth and frequency of tillage, as well as soil and climatic factors. Limiting tillage depth, speed and the number of operations all conserve crop residue and soil moisture. Shallow tillage allows crop residue to accumulate near the soil surface, where it will be most effective in reducing wind and water erosion, improving infiltration and reducing evaporation. Reducing tillage speed usually lessens crop residue burial.

Tillage equipment type significantly influences residue conservation. Tillage implements such as a wide blade cultivator or rod weeder conserve significantly more residue than a cultivator, which, in turn, is better than a discer. The addition of harrows to a field or heavy-duty cultivator doubles the amount of residue buried. The addition of a rod weeder to a cultivator does not significantly affect residue reduction.

Table: Approximate residue conservation with various tillage implements.

Tillage implement	Residue reduction per operation (%)	Residue left after four operations (%)
Wide blade cultivator (35 inch sweeps)	10	50-60
Rod weeder	5-10	no data
Heavy duty cultivator (16-18 inch sweeps)	20	30-40
Heavy duty cultivator with rodweeder	20	30-40
Heavy duty cultivator with harrows	40	15-20
Field cultivator (9-12 inch sweeps)	20	30-40
Field cultivator with harrows	40	15-20
Discer	35-65	10-15
Tanderm disc - offset disc	30-70	5-15
Moldboard plow	90	–

The need for each tillage operation should be carefully considered. For example, in the brown and dark brown soil zones fall tillage may be of little benefit unless a heavy infestation of moisture-consuming weeds is present. In many cases fall tillage reduces moisture conservation by disturbing standing stubble, which diminishes its effectiveness as a snow trap and exposes the soil to erosion. In the brown soil zone, standing stubble will conserve one-half to one inch more moisture over winter than bare soil. In contrast, fall tillage may offer some advantage in the more humid black and gray soil zones where moisture is rarely limiting and where crop residues may pose a problem during preparation of the seedbed. Even in these areas, however, fall tillage may not result in increased yields.

In all situations, avoid tillage under wet soil conditions as this can degrade soil structure and significantly reduce surface residue levels.

Stubble Cutting

Trapping more overwinter snow with "tall" or "sculptured" stubble enhances moisture conservation. Tall stubble refers to stubble that is cut 12 or more inches high, usually when straight combining. Sculptured stubble refers to when a swather-mounted clipper or deflector cuts strips of taller stubble with each swath, or when alternate swaths are cut at normal height and taller.

Direct Seeding

Zero-till, direct seeding is not usually associated with organic crop production because herbicides cannot be used. Organic producers, however, do practice direct seeding.

Thus, producers may wish to consider this conservation practice when fall and spring weed pressure is low, and previous crop straw and chaff have been adequately spread.

Extended Crop Rotations

Summerfallowing is destructive to the soil because no new organic matter is returned to the soil during the fallow year. Tillage also speeds the breakdown of soil organic matter and predisposes the soil to erosion. Extending crop rotations is a conservation practice because it reduces the incidence of summerfallow.

The benefits of an extended crop rotation are numerous and include improved fertility, tilth, aggregate stability, moisture storage, and resistance to soil erosion and degradation, as well as reductions in insect, weed and disease problems. All of these factors contribute to increased productivity and most have a significant positive effect on soil sustainability.

Decisions regarding cropping strategies should consider not only short-term benefits, but also their long-term effects on soil and environmental quality. A diverse rotation should include cereals, oilseeds, pulses, fall-seeded crops and forages. The level of crop diversity determines the significance and degree of the rotational benefits. Selection and management of legume species (pulses and forage legumes) within a rotation is a vital aspect of achieving diversity and supplying nitrogen through symbiotic nitrogen fixation.

Forage Crops

Forages contribute significant amounts of organic material to the soil. They also offer an alternative commodity in the form of hay, silage or seed. Semi-permanent forage production should be considered on soils, which are inherently low in productivity and/or vulnerable to erosion. Forage production for two to four years should also be considered as part of a normal crop rotation.

Selection of forage species and management practices can be tailored to specific problems such as drought, excessive soil moisture, salinity, poor soil structure, low pH and other problems.

Complementary Rotational Crops

An ideal rotation should be as diverse as possible including cereals, oilseeds, pulses, fall-seeded crops and forages. A diverse crop rotation, in addition to many other benefits, can help soil nutrient availability, because different crops have a different demand for and ability to remove particular nutrients.

All crops require 16 essential nutrients, most soils are deficient in nitrogen and available phosphorus. Some soils are also deficient in sulphur and potassium. Micronutrients are rarely deficient. For example, canola and alfalfa have a high demand for sulphur and should not be grown on deficient soils. Where copper is deficient, as in some grey soils, barley or rye will provide comparatively better yields than the other cereals.

Growing legumes in the rotation provides both nitrogen and non-nitrogen benefits to subsequent crops. Properly inoculated/nodulated legumes fix 50 to 90 percent of their nitrogen requirement from the air. The remainder is obtained from the soil. in addition, nitrogen is also exuded from legume roots during the growing season and can be utilized by the companion crop if intercrops are grown. The legume residue decomposes and recycles the nutrients faster than non-legume residues, thus more nitrogen is usually available to the subsequent crop than if a non-legume had been grown. Furthermore, research has shown that the non-nitrogen benefits (such as disease suppression, tilth improvement, etc.) of growing legumes in the rotation may result in increased yields.

The growth patterns of various crops should also be taken into account when planning complementary rotational cropping. Broadleaf crops such as pea, lentil, flax and polish canola generally extract moisture and nutrients from shallower depths than spring-seeded cereals. Fall rye and winter wheat root deeper, earlier in the growing season than spring cereals, using early spring moisture to their advantage. Winter cereals also have an advantage over spring cereals because they usually flower earlier, when soil moisture reserves are more plentiful.

Perennial forages can be very deep rooted, using moisture and nutrients from the subsoil. Shallow rooted crops appear to be the best adapted to follow a deep-rooted crop because water recharge is likely to occur only near the soil surface and a shallow-rooted crop will not expend energy rooting deeper in search of moisture that is not there. Medium or deep-rooted crops appear better adapted to follow shallow-rooted crops as they are able to take advantage of any moisture left at depth, not used by the previous shallow-rooted crop.

Wind Barriers

Annual Crop Barriers in Crop

Barriers of "taller" annual crops have been used to a limited degree in low residue-producing crops. A divider is placed in the seedbox so that two rows of wheat or flax are seeded every seeder width or two of lentil. At harvest, the lentil is combined and the barrier strip left standing to trap snow and prevent wind erosion during the upcoming winter.

Perennial Grass Barriers

Perennial grass barriers are two rows of grass planted perpendicular to prevailing winds to reduce wind erosion, trap snow and reduce evaporative losses. Barriers should be placed 30 to 60 feet apart, depending on soil type; closest on sands, moderately spaced on clays and furthest apart on loams. Barriers may be placed further apart if other soil conservation practices are also being used.

Species such as tall wheatgrass work well, as it is usually a weak competitor with most field crops and will not spread beyond the seeded rows. It also grows high enough without lodging to trap snow, helping in soil moisture recharge.

Shelterbelts

Shelterbelts can effectively reduce wind velocity for a distance approximately 20 or more times their height. This is usually sufficient to control wind erosion for a distance of approximately 10 or more times their height when planted perpendicular to prevailing winds. Their effectiveness, however, does depend upon shape, porosity and maintenance as well as height. Shelterbelts have the added benefit of increasing crop yields, particularly of less drought-tolerant crops such as canola and alfalfa.

Strip Cropping

Strip cropping consists of alternating strips of crop and summerfallow at an angle perpendicular to the prevailing winds. The strip width varies depending on soil texture. Sandy soils are the most prone to wind erosion, followed by clays, then loams. Strip cropping works well for loams to clays where eight to 10 strips per quarter section will significantly reduce the potential for wind erosion. With sandy textured soils, however, too many strips are required to be manageable. Keep field equipment sizes in mind when establishing strip widths.

Strip cropping is a more common practice in the drier areas of the prairies where often too little crop residue is present to prevent wind erosion. It can be used in wetter areas, however, provided the strips also run perpendicular to the slope, so that water erosion does not become a problem.

Cocktail Cover Crops

A cover crop mixture is one way to get the benefits of multiple cover crops, and on steep land, a high-residue cover crop to prevent erosion should be included in the cover crop mixture. Cereal crops should be included at a rate of one-third to one-half in the mixture and the others can be legume crops. Some of the cocktail cover crops will either be killed by fall frost remaining on the soil surface until spring planting, providing valuable soil protection or start growing the following spring. These crops may be removed by tillage or used for short term livestock grazing or grown to maturity in the case of winter wheat or fall rye.

References

- Managing-soil-fertility-organic-farming_system, soil-management-organic-farms, organic-crop-production: organic-crop-production.com, Retrieved 14 February, 2019

- Nutrientmanagementonorganicfarms: teagasc.ie, Retrieved 15 March, 2019

- Soil-fertility-in-organic-farming-systems:-much-more-than-plant-nutrition: extension.org, Retrieved 16 April, 2019

- Organic-crop-production-soil-conservation-practices, organic-crops, crops-and-irrigationm, agri-business-farmers-and-ranchers, agriculture-natural-resources-and-industry, business: saskatchewan.ca, Retrieved 17 May, 2019

5
Organic Pest and Disease Management

Organic pest and disease control includes diverse methods to manage pests and diseases such as biological pest control, organic weed control, cultural weed control and early blight management. This chapter discusses in detail these methods and practices related to organic pest and disease management.

Natural Pest and Disease Control

Pests and diseases are part of the natural environmental system. In this system there is a balance between predators and pests. This is nature's way of controlling populations. The creatures that we call pests and the organisms that cause disease only become 'pest and diseases' when their activities start to damage crops and affect yields. If the natural environmental system is imbalanced then one population can become dominant because it is not being preyed upon. The aim of natural control is to restore a balance between pest and predator and to keep pests and diseases down to an acceptable level. The aim is not to eradicate them altogether, as they also have a role to play in the natural system. Once a pest or disease has started to attack a crop, the damage cannot be repaired and control becomes increasingly difficult. Where possible, use techniques to avoid or prevent pest and disease attack in the first place.

Advantages of Natural Control

Pesticides do not solve the pest problem. In the past 50 years, insecticide use has increased tenfold, while crop losses from pest damage have doubled. Here are three important reasons why natural control is preferable to pesticide use.

Cost

Using natural pest and disease control is often cheaper than applying chemical pesticides because natural methods do not require buying expensive materials from the outside. Products and materials which are already in the home and around the farm are most often used.

Safety for People

There is much concern over the dangers of chemical products. They may be misused because the instructions are not written in the language spoken by the person using the product. There have been many reports of people suffering from severe skin rashes and headaches as a result of using chemical pesticides. There are an estimated one million cases of poisoning by pesticides each year around the world. Up to 20,000 of these result in death. Most of the deaths occur in developing countries where chemical pesticides, which are banned in Europe or the USA, are still available.

Safety for the Environment

Pests are often controlled with man made chemicals which have many harmful effects, for example:

- Artificial chemicals kill useful insects which eat pests.

- Artificial chemicals can stay in the environment and in the bodies of animals causing problems for many years.

- Artificial products are very simple chemicals and insect pests can very quickly, over a few breeding cycles, become resistant to them and can no longer be controlled.

Knowing the Problem

Before taking action to control pests and diseases it is very important to make sure that the problem is correctly identified. Only then can you hope to succeed. Knowledge of pests and diseases will help you to decide whether the problem is caused by a pest, a disease, a mineral deficiency in the soil or an environmental factor. A good identification book may help with this.

Proper identification should be the first step in controlling the problem and, more importantly, in preventing it from happening again.

The following pages describe a general approach to natural pest and disease control and give some specific examples.

A Healthy Soil

A soil managed using organic methods will give plants a balanced food supply. Plants which are fed well, like people, will be much more resistant to pest and disease. So caring for the soil is important. It should be managed in ways that develop and protect its structure, its fertility and also the millions of creatures for which it is a home. Caring for the soil involves providing a regular input of organic residues in the form of animal manures and plant remains. The aim is to:

- Maintain levels of humus (organic material) that give structure to the soil.

- Feed organisms which live in the soil.

- Provide nutrients for crops.

Whilst chemical fertilisers appear to improve plant growth, their use can also have negative effects. A plant may look healthy but, because of the high content of nitrogen given by the chemical fertiliser, causing fast sappy growth, it is very attractive to pests. It has been observed that aphids lay double the number of eggs on a plant grown with chemical fertilisers compared to organically grown plants.

A Healthy Crop

By giving plants the right growing conditions they will be more able to resist pests and diseases. Also, the right choice of crop will help to deter pests and disease. A crop growing in an area where it is not suited is more likely to be attacked. You should take account of the soil type, the climate, the altitude, the available nutrients and the amount of water needed when selecting your crops. Plants will only yield well and resist pests and diseases if they are grown under the most suitable conditions for that particular plant.

To help ensure a healthy crop, weeding should be done early and regularly to stop weeds from taking nutrients which should be going to the crop.

Resistant Varieties and Genetic Diversity

Within a single crop there can be many differences between plants. Some may be tall, some may be able to resist particular diseases.

There is most variety in the traditional crops grown by farmers. These have been grown and selected over many centuries to meet the requirements of the farmer. Although many of these are being replaced by modern varieties, seeds are often still saved locally.

Crops which have been bred by modern breeding methods tend to be very similar and if one plant is susceptible to a disease, all the other plants are as well. Although some new modern varieties may be very resistant to specific pests and diseases they are often less suited to the local climate and soil conditions than traditional varieties. It can therefore be dangerous to rely too much on any one of them.

Crop Rotation

Growing the same crops in the same site year after year can encourage a build up of pests and diseases in the soil. These will transfer from one crop to the next. Crops should be moved to a different area of land each year, and not returned to the original site for several years. For vegetables a 3 to 4 year rotation is usually recommended as a minimum.

Crop rotation also helps a variety of natural predators to survive on the farm.

A typical 4 year rotation would include a cycle with maize and beans, a cereal and a root crop with either of the following:

1. Grass or bush fallow (a fallow period where no crops are grown).

2. A legume crop where a green manure, which is a plant grown mainly for the benefit of the soil, is grown.

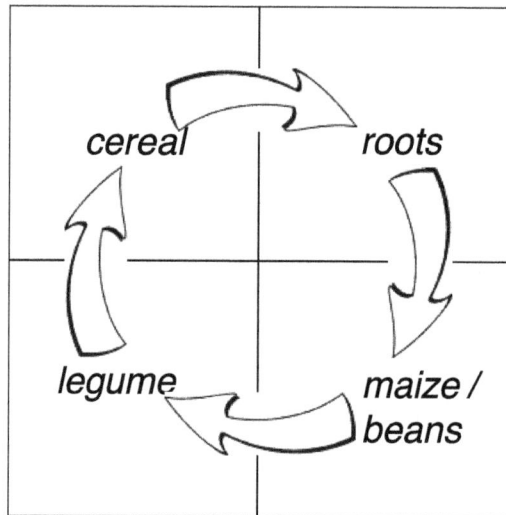

Crop rotation helps to control pests and diseases.

With crops such as brassicas and onions which are usually grown in a vegetable garden the whole year round, the populations of certain pests and diseases can keep increasing because there is always a suitable host plant for them. Breaking the cycle can help to solve the problem. This can be done through rotation within the vegetable garden.

Good Hygiene

If infected plant material, live or dead, is left lying around, pests and diseases can be passed on to future crops. Debris should be cleared up and disposed of. This can be done by composting the debris. The composting process will kill some pests and diseases and produce compost which is a good soil improver and fertiliser. Some diseases may survive being composted. If in doubt, the infected material should be burnt.

Soil Tillage

Many pests spend part of their lives as larvae or pupae in the soil. Ploughing or digging when the soil is dry can reveal the pest and they will dry out and die in the sun, or they can be picked off the ground by hand or birds or other predators. Ploughing can also push the pest deep down into the ground where they will not be able to survive. Ploughing and disturbing the soil should be carefully considered against the harmful effects it may have such as destroying the structure of the soil and causing soil erosion.

Soil pH

The pH (acidity or alkalinity) of a soil can affect some diseases. Changing the pH can reduce the problem.

For example, potato scab is less severe in more acid (pH below 7) soils. A layer of grass mowings added to the bottom of the potato trenches at planting time will make the soil more acid and reduce the disease.

Clubroot is less severe in alkaline conditions (pH above 7) therefore liming the soil to make it more alkaline can reduce the problem.

Timely Sowing

It is often the young of many pests (larvae, caterpillar), rather than the adults, that cause damage to crops. Problems can be avoided by delaying sowing until the egg laying period of a pest is over, or by protecting the plants during this period. It is therefore important to know the life cycle of pests, so that timely sowing can be carried out.

In Ghana, for example, farmers in the forest zone only plant maize in the main rainy season. In the lesser rainy season, the maize is attacked by stem borers.

Companion Planting

Companion planting means growing certain plants to protect other plants from pests or diseases. This may be because the pest is deterred by the companion plant, or because it is attracted to the companion plant rather than the crop.

For example onions planted either side of a row of carrots help to deter carrot flies. You need to sow 4 rows of onions for 1 row of carrots. This effect will only last as long as the onions are growing leaves. Many pests avoid garlic, so this can be used very effectively for companion planting with most crops.

In a similar way farmers in Zimbabwe have found that placing mint leaves near spinach plants will deter insect pests.

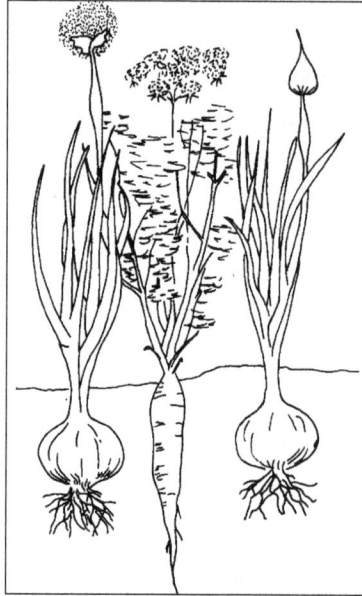

Growing onions with carrots to help deter carrotfly.

By planting milkweed among vegetables, some African farmers have effectively reduced the number of aphids on their crops. This is because aphids are more attracted to the milkweed than to the vegetables.

Companion planting can also mean that one plant acts as a barrier for another. In Columbia, jassid infestation in beans is reduced when beans are sown 20 to 30 days after maize. The maize acts as a shelter for the beans.

Plants to attract Predators and Parasites

Similarly to companion planting, which seeks to deter pests from the main crop, attractant plants can be grown to attract predatory insects.

Areas of Natural Habitat

Bushes and trees are a home for many useful insects and birds. They provide resting areas, shelter and food. Areas of natural habitat can be left around the edges of fields where crops are grown. If these areas are destroyed then there is likely to be an imbalance between the populations of predator and pest.

Specific Plants to attract Beneficial Insects

There are many plants that can be grown to attract natural predators and parasites which will help to keep down pests and diseases.

Flowers such as marigolds (Tagetes), mint (Mentha), sunflower (Helianthus annus), sunhemp (Crotalaria juncea) as well as local legumes are useful attractant plants.

Hoverflies, whose larvae feed on greenfly are attracted to the flowers of herbs and vegetables such as fennel, celery, dill, carrots and parsnips (Umbelliferae family). The nectar and pollen that these flowers provide will help to increase the number of eggs that these insects lay. Umbellifers will also provide food to various parasitic wasps whose young live on aphids and some caterpillars.

Red hot pokers (Kniphofia uvaria) are used in parts of Africa to attract birds that eat aphids.

Barriers

Barriers are physical structures put in place to prevent a pest from reaching a plant. They keep pests away from a plant but do not kill them. Here are some examples that you can adapt, depending on the resources available to you:

Crawling Insects

Cut the top off a transparent plastic bottle and place it firmly into the ground, over a young plant. This stops pests such as slugs from reaching the plant.

Using an old plastic bottle to protect a young plant.

Climbing Insects

To help protect trees from attack by insects, grease bands can be used. Wrap a piece of plastic or a long leaf around the trunk of the tree. Spread any kind of thick grease on top of this. Fold over the top of the foil or plastic to form an overhang to protect the grease from being washed away by rain. Check the grease every week to ensure that the grease is intact. This prevents crawling insects such as ants, fruit fly larvae, slugs, snails, beetles or caterpillars from damaging trees, especially fruit trees, or grain stores.

Termites

Digging a 70-100 cm trench around buildings and nurseries can prevent attack from subterranean species of termites. This is a good method of control however it is hard work. Alternatively, barriers can be built. These should be partially above and below ground and should be made from material that is impenetrable to termites such as basalt, sand or crushed volcanic cinders. Particle size of the material is critical, they should not be too large for the termites to carry away, and not so small that termites can pack the particles to create a continuous passage through which they can move.

Bait Traps

The use of baits and traps are traditional methods which have become neglected because of the increasing use of chemical pesticides. Here are some examples:

Cutworms

Method One

Mix equal quantities of hardwood sawdust, bran, molasses and enough water to make the solution sticky. Spread around the base of the plants in the evenings. The molasses attract the cutworms and as they try to pass through it they get stuck. The substance dries out in the sun and the pest dies.

Method Two

Mix 100 grams (g) of bran, 10 g of sugar, 200 g of water, 5 g of pyrethrum powder. Spread around the base of the plants. The cutworms eat the substance and die.

Spread bait around the base of plants to trap cutworms.

Fruit Fly

Traps need to be put in place before an attack is likely to start. For fruit fly, the traps should be baited 6 to 8 weeks before the fruit ripen.

Here are two examples of trap constructions which could be adapted:

Method One

Make a small hole in the bottom of a plastic bottle or container. Seal the top of the bottle with a lid or stopper. Fill one quarter of the bottle with the bait. Hang the bottle upside down from trees around fields or gardens. The flies are attracted to the bait through the small hole. They are then trapped and drown in the bait.

Method Two

Cut the top of a plastic bottle off. Pour some bait in the bottom half of the bottle. Turn the top half of the bottle upside down and place in the bottom half. Again the flies are attracted into the bottle and drown in the bait.

Here are two different baits for fruit fly that can be poured into the traps:

- Mix 1 litre of water, 250 millilitres (ml) of urine, a few drops of vanilla essence, 100 g of sugar and 10 g of pyrethrum powder.

- Mix 1 teaspoon of pyrethrum powder, 250 g of honey, a few drops of vanilla essence, 250 g of orange or cucumber peel or pulp and 10 litres of water.

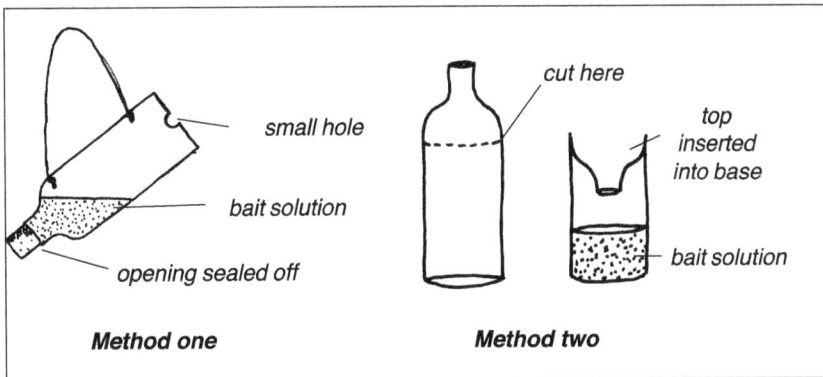

Examples of fruit fly traps filled with bait.

Light Traps

Light traps are set up at night and attract a variety of flying insects including moths, mosquitoes, chafer beetles, american bollworms, army worms, cutworms, brown rice plant hopper, green rice leaf hopper, rice black bugs, rice gall midges, rice stem borers and tomato hornworms.

Make a tripod construction from wooden poles or bamboo. Press the poles down firmly into the ground to secure it so that it cannot be blown over or knocked down by animals. Suspend a lantern from the top of the construction over a bowl of water with a little oil in it.

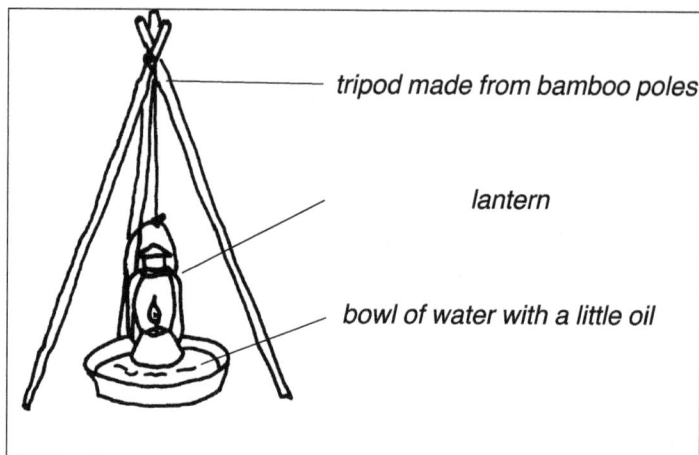

tripod made from bamboo poles

lantern

bowl of water with a little oil

Example of a light trap.

Fire risks must be kept in mind and the lamps must be hung so that the wood does not catch fire.

The best timing for placing light traps around a garden or field, depends on the life cycle of the insect and the development stage of the crop. The best time is just after the moths emerge but before they lay eggs, so it is important to know the life cycle of the pests.

Fly Trap

Fly traps are large boards measuring about 30 cm by 30 cm which are painted bright yellow/orange and covered with an adhesive such as oil or glue. Different pests are attracted to different colours so you need to experiment. The flies are attracted to the bright colour of the board and fly onto it. They get stuck in the oil or glue and die.

For example, leaf minors are attracted to yellow, so place several yellow boards 60 cm off the ground (on a table or hung from a tree). The board will attract a huge number of insects, which means a considerable reduction of pests.

Pheromone Traps

Pheromone is the sexual attractant produced by some female insects. If a trap is baited with this it will attract the male insects into the trap from which they cannot escape. Pheromone traps alone can reduce pest damage. Alternatively, they give an indication of pest populations and therefore the best time to apply control methods. Pheromones traps are usually prepared by commercial companies and may be costly to the farmer.

However, if you have a particularly severe pest problem it may be worth investing in one rather than using chemical pesticides.

Hand Picking and Squashing

In some cases it may be possible to pick pests directly off the crops. This can be done especially with caterpillars and other large insects in small plots of land. Smaller pests such as aphids can be squashed on the plant.

Parts of plants that are diseased can be cut or broken off the plants to prevent the spread of the disease.

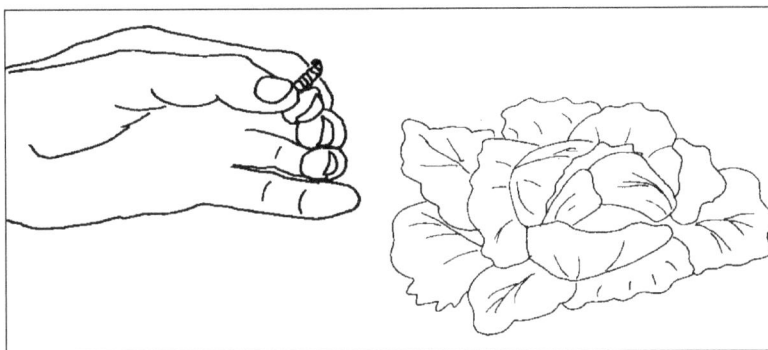

Handpicking caterpillars from a cabbage plant.

Biological Control

Biological control means using one creature or organism to control a pest. This often involves introducing a creature or organism, which is known to be predatory, to an area with the aim that it will control the population of the pest.

Some widespread pest and disease problems have been dealt with in this way by government projects. For example, a variety of creatures have been introduced to control the cassava mealy bug in Kenya. Here are other examples of creatures or organisms which are known to control certain pests:

- Control of Cabbage caterpillars: Bacillus thuringiensis is a bacteria which kills many types of caterpillar, but only when they eat it. This bacteria (which can be bought as a commercial product called "Bactospeine") is applied to brassicas (cauliflowers, cabbages) as a spray.

- Control of Vine weevils: Nemasys H is a preparation containing parasitic nematodes which seek out and destroy vine weevil larvae. It is watered onto the soil.

Biological control does not have to involve buying commercial products. It can be achieved on a small-scale by encouraging natural predators to live and breed in the area where pests are a problem. This can be achieved by having trees and hedges around the farm to provide a home for them. There are many insects and animals which should be encouraged because they feed on pests. Here are some examples:

Frogs, toads, hedgehogs, mice, moles, bats, birds, chamelions, lizards, spiders, ants, assasin bugs, black-kneed capsids, bees, branchid wasps, parasitic wasps, dung beetles, ground beetles, earthworms, hawk moths, dragon flies, hoverflies, lacewings and stick insects.

Natural Pesticides

If pests and diseases cannot be prevented or controlled by cultural and physical means, it may be necessary to use natural pesticides. Many growers have developed ways of making their own sprays from plants such as garlic, chillies, marigolds and many others. These are inexpensive and have proved to be very effective. Here are some examples:

A solution can be made from marigold using water and soap. The liquid acts as a crop strengthener to help potatoes, beans, tomatoes and peas resist blight, mildew and other fungal diseases. It also repels aphids, caterpillars and flies.

Garlic spray is particularly good against army worms, Colorado Beetle, False codling moth, Khapra beetle, Mexican bean beetle and Imported cabbage worm. Garlic can also kill nematodes if soil or batches of soil are drenched with garlic liquid.

Biological Pest Control

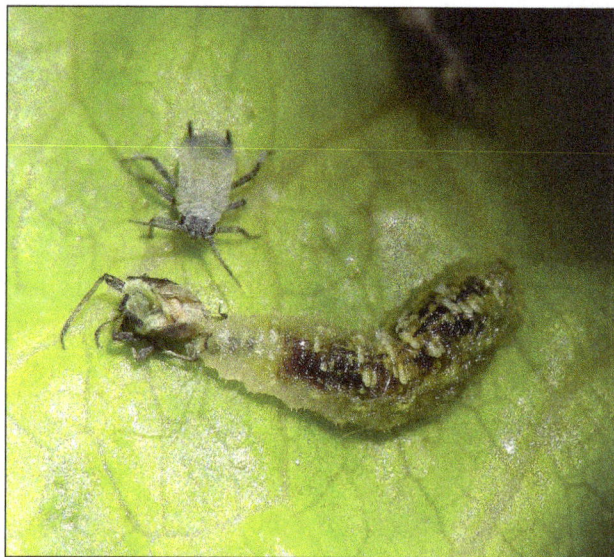

Syrphus hoverfly larva (below) feed on aphids (above), making them natural biological control agents.

A parasitoid wasp (*Cotesia congregata*) adult with pupal cocoons on its host, a tobacco hornworm (*Manduca sexta*, green background), an example of a hymenopteran biological control agent.

Biological control or biocontrol is a method of controlling pests such as insects mites, weeds and plant diseasesusing other organisms. It relies on predation, parasitism, herbivory, or other natural mechanisms, but typically also involves an active human management role. It can be an important component of integrated pest management (IPM) programs.

There are three basic strategies for biological pest control: classical (importation), where a natural enemy of a pest is introduced in the hope of achieving control; inductive (augmentation), in which a large population of natural enemies are administered for quick pest control; and inoculative (conservation), in which measures are taken to maintain natural enemies through regular reestablishment.

Natural enemies of insect pests, also known as biological control agents, include predators, parasitoids, pathogens, and competitors. Biological control agents of plant diseases are most often referred to as antagonists. Biological control agents of weeds include seed predators, herbivores and plant pathogens.

Biological control can have side-effects on biodiversitythrough attacks on non-target species by any of the same mechanisms, especially when a species is introduced without thorough understanding of the possible consequences.

Types of Biological Pest Control

There are three basic biological pest control strategies: importation (classical biological control), augmentation and conservation.

Importation

Importation or classical biological control involves the introduction of a pest's natural enemies to a new locale where they do not occur naturally. Early instances were often unofficial and not based on research, and some introduced species became serious pests themselves.

Rodolia cardinalis, the vedalia beetle, was imported from Australia to California successfully controlling cottony cushion scale.

To be most effective at controlling a pest, a biological control agent requires a colonizing ability which allows it to keep pace with changes to the habitat in space and time. Control is greatest if the agent has temporal persistence, so that it can maintain its population even in the temporary absence of the target species, and if it is an opportunistic forager, enabling it to rapidly exploit a pest population.

Joseph Needham noted a Chinese text dating from 304 AD, *Records of the Plants and Trees of the Southern Regions*, by Hsi Han, which describes mandarin oranges protected by large reddish-yellow citrus ants which attack and kill insect pests of the orange trees. The citrus ant (*Oecophylla smaragdina*) was rediscovered in the 20th century, and since 1958 has been used in China to protect orange groves.

One of the earliest successes in the west was in controlling *Icerya purchasi* (cottony cushion scale) in Australia, using a predatory insect *Rodolia cardinalis* (the vedalia beetle). This success was repeated in California using the beetle and a parasitoidal fly, *Cryptochaetum iceryae*. Other successful cases include the control of *Antonina graminis* in Texas by *Neodusmetia sangwani* in the 1960s.

Damage from *Hypera postica*, the alfalfa weevil, a serious introduced pest of forage, was substantially reduced by the introduction of natural enemies. 20 years after their introduction the population of weevilsin the alfalfa area treated for alfalfa weevil in the Northeastern United States remained 75 percent down.

Alligator weed was introduced to the United States from South America. It takes root in shallow water, interfering with navigation, irrigation, and flood control. The alligator weed flea beetle and two other biological controls were released in Florida, greatly reducing the amount of land covered by the plant.Another aquatic weed, the giant salvinia (*Salvinia molesta*) is a serious pest, covering waterways, reducing water flow and harming native species. Control with the salvinia weevil (*Cyrtobagous salviniae*) and the salvinia stem-borer moth (*Samea multiplicalis)* is effective in warm climates, and in Zimbabwe, a 99% control of the weed was obtained over a two-year period.

The invasive species *Alternanthera philoxeroides* (alligator weed) was controlled in Florida (U.S.) by introducing alligator weed flea beetle.

Small commercially reared parasitoidal wasps, *Trichogramma ostriniae*, provide limited and erratic control of the European corn borer (*Ostrinia nubilalis*), a serious pest. Careful formulations of the bacterium *Bacillus thuringiensis* are more effective.

The population of *Levuana iridescens*, the Levuana moth, a serious coconut pest in Fiji, was brought under control by a classical biological control program in the 1920s.

Augmentation

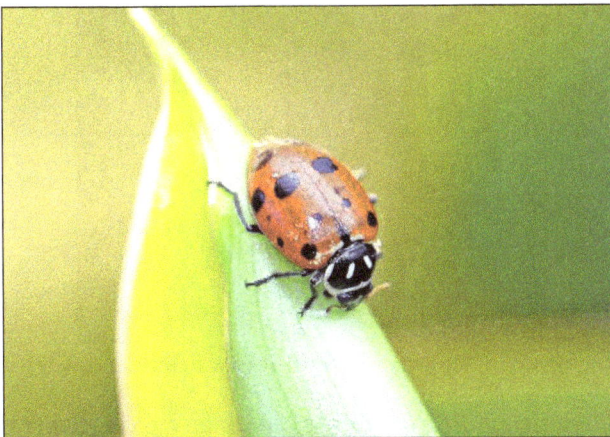

Augmentation involves the supplemental release of natural enemies that occur in a particular area, boosting the naturally occurring populations there. In inoculative release, small numbers of the control agents are released at intervals to allow them to reproduce, in the hope of setting up longer-term control, and thus keeping the pest down to a low level, constituting prevention rather than cure. In inundative release, in contrast, large numbers are released in the hope of rapidly reducing a damaging pest population, correcting a problem that has already arisen. Augmentation can be effective, but is not guaranteed to work, and depends on the precise details of the interactions between each pest and control agent.

An example of inoculative release occurs in the horticultural production of several crops in greenhouses. Periodic releases of the parasitoidal wasp, *Encarsia formosa*, are used to control greenhouse whitefly,while the predatory mite *Phytoseiulus persimilis* is used for control of the two-spotted spider mite.

The egg parasite *Trichogramma* is frequently released inundatively to control harmful moths. Similarly, *Bacillus thuringiensis* and other microbial insecticides are used in large enough quantities for a rapid effect. Recommended release rates for *Trichogramma* in vegetable or field crops range from 5,000 to 200,000 per acre (1 to 50 per square metre) per week according to the level of pest infestation. Similarly, nematodes that kill insects (that are entomopathogenic) are released at rates of millions and even billions per acre for control of certain soil-dwelling insect pests.

Conservation

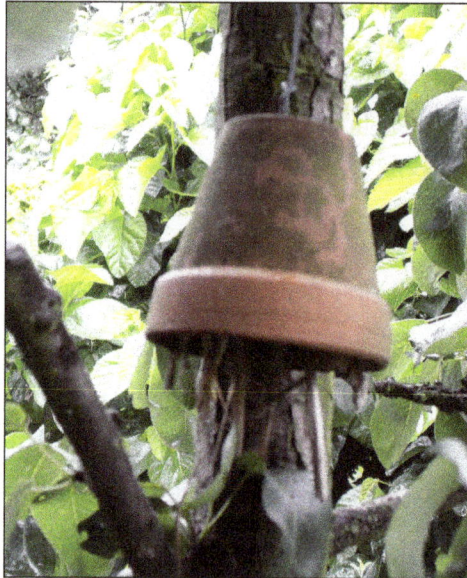

An inverted flowerpot filled with straw to attract earwigs.

The conservation of existing natural enemies in an environment is the third method of biological pest control. Natural enemies are already adapted to the habitat and to the target pest, and their conservation can be simple and cost-effective, as when nectar-producing crop plants are grown in the borders of rice fields. These provide nectar to support parasitoids and predators of planthopper pests and have been demonstrated to be so effective (reducing pest densities by 10- or even 100-fold) that farmers sprayed

70% less insecticides and enjoyed yields boosted by 5%. Predators of aphids were similarly found to be present in tussock grasses by field boundary hedges in England, but they spread too slowly to reach the centres of fields. Control was improved by planting a metre-wide strip of tussock grasses in field centres, enabling aphid predators to overwinter there.

Cropping systems can be modified to favor natural enemies, a practice sometimes referred to as habitat manipulation. Providing a suitable habitat, such as a shelterbelt, hedgerow, or beetle bank where beneficial insects such as parasitoidal wasps can live and reproduce, can help ensure the survival of populations of natural enemies. Things as simple as leaving a layer of fallen leaves or mulch in place provides a suitable food source for worms and provides a shelter for insects, in turn being a food source for such beneficial mammals as hedgehogs and shrews. Compost piles and stacks of wood can provide shelter for invertebrates and small mammals. Long grass and pondssupport amphibians. Not removing dead annuals and non-hardy plants in the autumn allows insects to make use of their hollow stems during winter. In California, prune trees are sometimes planted in grape vineyards to provide an improved overwintering habitat or refuge for a key grape pest parasitoid. The providing of artificial shelters in the form of wooden caskets, boxes or flowerpots is also sometimes undertaken, particularly in gardens, to make a cropped area more attractive to natural enemies. For example, earwigs are natural predators which can be encouraged in gardens by hanging upside-down flowerpots filled with straw or wood wool. Green lacewings can be encouraged by using plastic bottles with an open bottom and a roll of cardboard inside. Birdhouses enable insectivorous birds to nest; the most useful birds can be attracted by choosing an opening just large enough for the desired species.

In cotton production, the replacement of broad-spectrum insecticides with selective control measures such as Bt cotton can create a more favorable environment for natural enemies of cotton pests due to reduced insecticide exposure risk. Such predators or parasitoids can control pests not affected by the Bt protein. Reduced prey quality and abundance associated increased control from Bt cotton can also indirectly decrease natural enemy populations in some cases, but the percentage of pests eaten or parasitized in Bt and non-Bt cotton are often similar.

Biological Control Agents

Predators

Predators are mainly free-living species that directly consume a large number of prey during their whole lifetime. Given that many major crop pests are insects, many of the predators used in biological control are insectivorous species. Lady beetles, and in particular their larvae which are active between May and July in the northern hemisphere, are voracious predators of aphids, and also consume mites, scale insects and small caterpillars. The spotted lady beetle (*Coleomegilla maculata*) is also able to feed on the eggs and larvae of the Colorado potato beetle(*Leptinotarsa decemlineata*).

Predatory lacewings are available from biocontrol dealers.

The larvae of many hoverfly species principally feed upon aphids, one larva devouring up to 400 in its lifetime. Their effectiveness in commercial crops has not been studied.

Predatory *Polistes* wasp searching for bollworms or other caterpillars on a plant.

Several species of entomopathogenic nematode are important predators of insect and other invertebrate pests. *Phasmarhabditis hermaphrodita* is a microscopic nematode that kills slugs. Its complex life cycle includes a free-living, infective stage in the soil where it becomes associated with a pathogenic bacteria such as *Moraxella osloensis*. The nematode enters the slug through the posterior mantle region, thereafter feeding and reproducing inside, but it is the bacteria that kill the slug. The nematode is available commercially in Europe and is applied by watering onto moist soil.

Species used to control spider mites include the predatory mites *Phytoseiulus persimilis, Neoseilus californicus,* and *Amblyseius cucumeris*, the predatory midge *Feltiella acarisuga*, and a ladybird *Stethorus punctillum*. The bug *Orius insidiosus* has been successfully used against the two-spotted spider mite and the western flower thrips (*Frankliniella occidentalis*).

Predators including *Cactoblastis cactorum* (mentioned above) can also be used to destroy invasive plant species. As another example, the poison hemlock moth (*Agonopterix alstroemeriana)* can be used to control poison hemlock (*Conium maculatum*). During its larval stage, the moth strictly consumes its host plant, poison hemlock, and can exist at hundreds of larvae per individual host plant, destroying large swathes of the hemlock.

The parasitoid wasp *Aleiodes indiscretus* parasitizing a gypsy mothcaterpillar, a serious pest of forestry.

For rodent pests, cats are effective biological control when used in conjunction with reduction of "harborage"/hidinglocations. While cats are effective at preventing rodent "population explosions", they are not effective for eliminating pre-existing severe infestations. Barn owls are also sometimes used as biological rodent control. Although there are no quantitative studies of the effectiveness of barn owls for this purpose, they are known rodent predators that can be used in addition to or instead of cats; they can be encouraged into an area with nest boxes.

Parasitoids

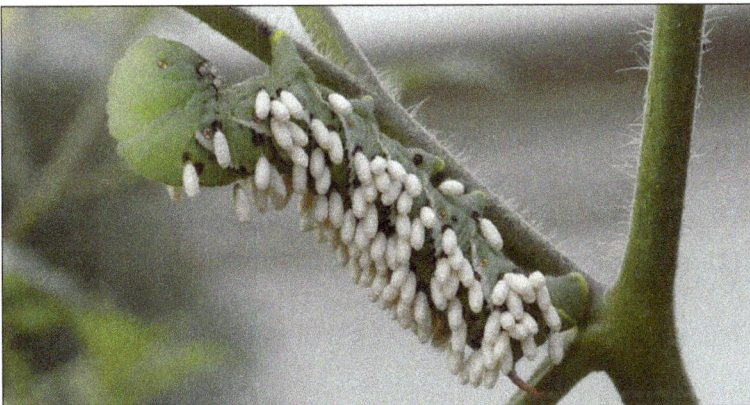

Parasitoids lay their eggs on or in the body of an insect host.

Parasitoids lay their eggs on or in the body of an insect host, which is then used as a food for developing larvae. The host is ultimately killed. Most insect parasitoid-sare wasps or flies, and many have a very narrow host range. The most important groups are the ichneumonid wasps, which mainly use caterpillars as hosts; braconid wasps, which attack caterpillars and a wide range of other insects including aphids; chalcid wasps, which parasitize eggs and larvae of many insect species; and tachinid flies, which parasitize a wide range of insects including caterpillars, beetleadults and larvae, and true bugs. Parasitoids are most effective at reducing pest populations when their host organisms have limited refuges to hide from them.

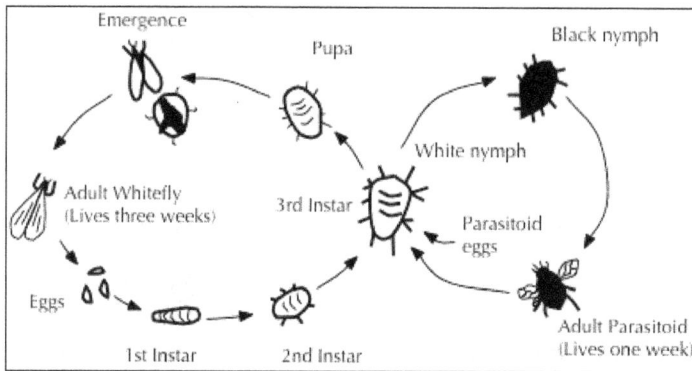

Life cycles of greenhouse whitefly and its parasitoid wasp *Encarsia formosa*.

Parasitoids are among the most widely used biological control agents. Commercially, there are two types of rearing systems: short-term daily output with high production of parasitoids per day, and long-term, low daily output systems. In most instances, production will need to be matched with the appropriate release dates when susceptible host species at a suitable phase of development will be available. Larger production facilities produce on a yearlong basis, whereas some facilities produce only seasonally. Rearing facilities are usually a significant distance from where the agents are to be used in the field, and transporting the parasitoids from the point of production to the point of use can pose problems. Shipping conditions can be too hot, and even vibrations from planes or trucks can adversely affect parasitoids.

Encarsia formosa is a small predatory chalcid wasp which is a parasitoid of white-fly, a sap-feeding insect which can cause wilting and black sooty moulds in glasshouse vegetable and ornamental crops. It is most effective when dealing with low level infestations, giving protection over a long period of time. The wasp lays its eggs in young whitefly 'scales', turning them black as the parasite larvae pupate. *Gonatocerus ashmeadi* (Hymenoptera: Mymaridae) has been introduced to control the glassy-winged sharpshooter *Homalodisca vitripennis* (Hemiptera: Cicadellidae) in French Polynesia and has successfully controlled ~95% of the pest density.

The eastern spruce budworm is an example of a destructive insect in fir and spruce forests. Birds are a natural form of biological control, but the *Trichogramma minutum*, a species of parasitic wasp, has been investigated as an alternative to more controversial chemical controls.

There are a number of recent studies pursuing sustainable methods for controlling urban cockroaches using parasitic wasps. Since most cockroaches remain in the sewer system and sheltered areas which are inaccessible to insecticides, employing active-hunter wasps is a strategy to try and reduce their populations.

Pathogens

Pathogenic micro-organisms include bacteria, fungi, and viruses. They kill or debilitate their host and are relatively host-specific. Various microbial insect diseases occur naturally, but may also be used as biological pesticides. When naturally occurring, these outbreaks are density-dependent in that they generally only occur as insect populations become denser.

Bacteria

Bacteria used for biological control infect insects via their digestive tracts, so they offer only limited options for controlling insects with sucking mouth parts such as aphids and scale insects. *Bacillus thuringiensis*, a soil-dwelling bacterium, is the most widely applied species of bacteria used for biological control, with at least four sub-species used against Lepidopteran (moth, butterfly), Coleopteran (beetle) and Dipteran (true fly) insect pests. The bacterium is available to organic farmers in sachets of dried spores which are mixed with water and sprayed onto vulnerable plants such as brassicas and fruit trees. Genes from *B. thuringiensis* have also been incorporated into transgenic crops, making the plants express some of the bacterium's toxins, which are proteins. These confer resistance to insect pests and thus reduce the necessity for pesticide use. If pests develop resistance to the toxins in these crops, *B. thuringiensis* will become useless in organic farming also. The bacterium *Paenibacillus popilliae* which causes milky spore disease has been found useful in the control of Japanese beetle, killing the larvae. It is very specific to its host species and is harmless to vertebrates and other invertebrates.

Fungi

Entomopathogenic fungi, which cause disease in insects, include at least 14 species that attack aphids. *Beauveria bassiana* is mass-produced and used to manage a wide variety of insect pests including whiteflies, thrips, aphids and weevils. *Lecanicillium* spp. are deployed against white flies, thrips and aphids. *Metarhizium* spp. are used against pests including beetles, locusts and other grasshoppers, Hemiptera, and spider mites. *Paecilomyces fumosoroseus* is effective against white flies, thrips and aphids; *Purpureocillium lilacinus* is used against root-knot nematodes, and 89 *Trichoderma* species against certain plant pathogens. *Trichoderma viride* has been used against Dutch elm disease, and has shown some effect in suppressing silver leaf, a disease of stone fruits caused by the pathogenic fungus *Chondrostereum purpureum*.

Green peach aphid, a pest in its own right and a vector of plant viruses, killed
by the fungus *Pandoraneoaphidis* (Zygomycota: Entomophthorales).

The fungi *Cordyceps* and *Metacordyceps* are deployed against a wide spectrum of arthropods. *Entomophaga* is effective against pests such as the green peach aphid.

Several members of Chytridiomycota and Blastocladiomycota have been explored as agents of biological control. From Chytridiomycota, *Synchytrium solstitiale* is being considered as a control agent of the yellow star thistle (*Centaurea solstitialis*) in the United States.

Viruses

Baculoviruses are specific to individual insect host species and have been shown to be useful in biological pest control. For example, the Lymantria dispar multicapsid nuclear polyhedrosis virus has been used to spray large areas of forest in North America where larvae of the gypsy moth are causing serious defoliation. The moth larvae are killed by the virus they have eaten and die, the disintegrating cadavers leaving virus particles on the foliage to infect other larvae.

A mammalian virus, the rabbit haemorrhagic disease virus was introduced to Australia to attempt to control the European rabbit populations there. It escaped from quarantine and spread across the country, killing large numbers of rabbits. Very young animals survived, passing immunity to their offspring in due course and eventually producing a virus-resistant population. Introduction into New Zealand in the 1990s was similarly successful at first, but a decade later, immunity had developed and populations had returned to pre-RHD levels.

Oomycota

Lagenidium giganteum is a water-borne mould that parasitizes the larval stage of mosquitoes. When applied to water, the motile spores avoid unsuitable host species and search out suitable mosquito larval hosts. This mould has the advantages of a dormant phase, resistant to desiccation, with slow-release characteristics over several years. Unfortunately, it is susceptible to many chemicals used in mosquito abatement programmes.

Competitors

The legume vine *Mucuna pruriens* is used in the countries of Benin and Vietnam as a biological control for problematic *Imperata cylindrica* grass: the vine is extremely vigorous and suppresses neighbouring plants by out-competing them for space and light. *Mucuna pruriens* is said not to be invasive outside its cultivated area. *Desmodium uncinatum* can be used in push-pull farming to stop the parasitic plant, witchweed (*Striga*).

The Australian bush fly, *Musca vetustissima*, is a major nuisance pest in Australia, but native decomposers found in Australia are not adapted to feeding on cow dung, which is where bush flies breed. Therefore, the Australian Dung Beetle Project (1965–1985), led by George Bornemissza of the Commonwealth Scientific and Industrial Research Organisation, released forty-nine species of dung beetle, to reduce the amount of dung and therefore also the potential breeding sites of the fly.

Combined use of Parasitoids and Pathogens

In cases of massive and severe infection of invasive pests, techniques of pest control are often used in combination. An example is the emerald ash borer, *Agrilus planipennis*, an invasive beetle from China, which has destroyed tens of millions of ash trees in its introduced range in North America. As part of the campaign against it, from 2003 American scientists and the Chinese Academy of Forestry searched for its natural enemies in the wild, leading to the discovery of several parasitoid wasps, namely *Tetrastichus planipennisi*, a gregarious larval endoparasitoid, *Oobius agrili*, a solitary, parthenogenic egg parasitoid, and *Spathius agrili*, a gregarious larval ectoparasitoid. These have been introduced and released into the United States of America as a possible biological control of the emerald ash borer. Initial results for *Tetrastichus planipennisi* have shown promise, and it is now being released along with *Beauveria bassiana*, a fungal pathogen with known insecticidal properties.

Difficulties

Many of the most important pests are exotic, invasive species that severely impact agriculture, horticulture, forestry and urban environments. They tend to arrive without their co-evolved parasites, pathogens and predators, and by escaping from these, populations may soar. Importing the natural enemies of these pests may seem a logical move but this may have unintended consequences; regulations may be ineffective and there may be unanticipated effects on biodiversity, and the adoption of the techniques may prove challenging because of a lack of knowledge among farmers and growers.

Side Effects

Biological control can affect biodiversity through predation, parasitism, pathogenicity, competition, or other attacks on non-target species. An introduced control does not always target only the intended pest species; it can also target native species. In Hawaii during the 1940s parasitic wasps were introduced to control a lepidopteran pest and the wasps are still found there today. This may have a negative impact on the native ecosystem; however, host range and impacts need to be studied before declaring their impact on the environment.

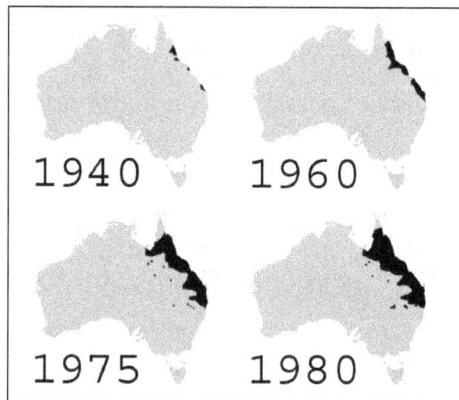

Cane toad (introduced into Australia 1935) spread from 1940 to 1980: it was ineffective as a control agent. Its distribution has continued to widen since 1980.

Vertebrate animals tend to be generalist feeders, and seldom make good biological control agents; many of the classic cases of "biocontrol gone awry" involve vertebrates. For example, the cane toad (*Rhinella marina*) was intentionally introduced to Australia to control the greyback cane beetle (*Dermolepida albohirtum*), and other pests of sugar cane. 102 toads were obtained from Hawaii and bred in captivity to increase their numbers until they were released into the sugar cane fields of the tropic north in 1935. It was later discovered that the toads could not jump very high and so were unable to eat the cane beetles which stayed on the upper stalks of the cane plants. However, the toad thrived by feeding on other insects and soon spread very rapidly; it took over native amphibian habitat and brought foreign disease to native toads and frogs, dramatically reducing their populations. Also, when it is threatened or handled, the cane toad releases poison from parotoid glands on its shoulders; native Australian species such as goannas, tiger snakes, dingos and northern quolls that attempted to eat the toad were harmed or killed. However, there has been some recent evidence that native predators are adapting, both physiologically and through changing their behaviour, so in the long run, their populations may recover.

Rhinocyllus conicus, a seed-feeding weevil, was introduced to North America to control exotic musk thistle (*Carduus nutans*) and Canadian thistle (*Cirsium arvense*). However, the weevil also attacks native thistles, harming such species as the endemic Platte thistle (*Cirsium neomexicanum*) by selecting larger plants (which reduced the gene pool), reducing seed production and ultimately threatening the species' survival. Similarly, the weevil *Larinus planus* was also used to try to control the Canadian thistle, but it damaged other thistles as well. This included one species classified as threatened.

The small Asian mongoose (*Herpestus javanicus*) was introduced to Hawaii in order to control the ratpopulation. However, the mongoose was diurnal, and the rats emerged at night; the mongoose therefore preyed on the endemic birds of Hawaii, especially their eggs, more often than it ate the rats, and now both rats and mongooses threaten the birds. This introduction was undertaken without understanding the consequences of such an action. No regulations existed at the time, and more careful evaluation should prevent such releases now.

The sturdy and prolific eastern mosquitofish (*Gambusia holbrooki*) is a native of the southeastern United States and was introduced around the world in the 1930s and '40s to feed on mosquito larvae and thus combat malaria. However, it has thrived at the expense of local species, causing a decline of endemic fish and frogs through competition for food resources, as well as through eating their eggs and larvae. In Australia, control of the mosquitofish is the subject of discussion; in 1989 researchers A. H. Arthington and L. L. Lloyd stated that "biological population control is well beyond present capabilities".

Grower Education

A potential obstacle to the adoption of biological pest control measures is that growers may prefer to stay with the familiar use of pesticides. However, pesticides have undesired effects, including the development of resistance among pests, and the destruction of natural enemies; these may in turn enable outbreaks of pests of other species than the ones originally targeted, and on crops at a distance from those treated with pesticides. One method of increasing grower adoption of biocontrol methods involves letting them learn by doing, for example showing them simple field experiments, enabling them to observe the live predation of pests, or demonstrations of parasitised pests. In the Philippines, early season sprays against leaf folder caterpillars were common practice, but growers were asked to follow a 'rule of thumb' of not spraying against leaf folders for the first 30 days after transplanting; participation in this resulted in a reduction of insecticide use by 1/3 and a change in grower perception of insecticide use.

Organic Weed Control

Weeds compete with crops for nutrients, light and water. This is why it is important to keep them in check. Growing organically allows you to manage weeds creatively and effectively, by careful planning of your planting and using mulches. Not all weeds are bad - some attract pollinators and some can improve the soil. And of course, most weeds can be composted, which in turn will add nutrients to your soil.

Don't use toxic weed killers. They are bad for the environment and bad for you.

- It is a toxic herbicide used to kill weeds.

- On its own, glyphosate has limited toxicity. However, it is commonly mixed in chemical formulations to have maximum effect. These formulations, such as Roundup or Weedol Path Clear, are potentially far more dangerous.

- The European Food Safety Authority (EFSA) says that glyphosate is safe. However, most of their research is industry led, and they haven't tested the various individual commercial formulations.

- Independent research indicates that glyphosate is not only probably carcinogenic, but that it also affects the body's endocrine system – causing problems in the liver and kidneys. Industry testers dispute this, but have declined to reveal all the results of their safety tests.

- Over 30% of the bread in the UK contains traces of glyphosate. While not necessarily toxic in small amounts, this gradual and persistent intake could create a health risk.

Get Rid of Weeds without Toxic Chemicals

Patios/paths and other hard surfaces:

- Use boiling water, or a sharp knife or trowel.

- Flame or thermal weeders are effective, especially on young weeds, and relatively cheap to buy.

- Organic weedkiller is usually based on pelargonic acid - which kills the foliage, but doesn't penetrate the root.

- Vinegar is not recommended. Some local councils use DEFRA approved acetic acid (the basis of vinegar). As this acid is a chemical compound it is not organic. Domestic vinegar from a bottle has not been proved effective. Bleach is also not recommended.

- When creating a new path or patio, make an impenetrable foundation layer: a geotextile membrane or a substantial mix of hardcore rubble and sand, firmly flattened to exclude all light and reduce moisture.

- Cracks between pavers should be filled with mortar - not sand, which provides the ideal medium for weeds to germinate. Lime mortars are more environmentally friendly than cement mixes.

Larger areas, such as an overgrown allotment or veg patch:

- To clear an overgrown growing area, slash down the high standing weeds and then cover with a thick compost manure mulch (at least 20 cms) and/or a plastic sheet. Without light the weeds will weaken and eventually die off. Use the slashed foliage and stems on the compost heap.

- Dig up deep rooted weeds, such as dandelions and docks. Put foliage on the compost heap, and drown the roots in a bucket of water for a month or so. The water can be used as a liquid feed.

- Persistent weeds such as bindweed and ground elder have to be dug over regularly, removing as much root as possible. Every little bit. In some cases, it is worth digging up individual plants which you want to keep, cleaning their roots of the weed's root fragments, and then replacing in the bed which has also been dug over. Persistent digging, removing roots, will eventually – perhaps over a couple of years – weaken the plant and make it easier to keep on top of. Put foliage and roots (but not flowers or seed heads) into a black plastic sack. Tie up the top and leave in an out-of-the-way corner until it turns into gooey sludge, then compost it.

Remember, it is not just your own health that will benefit from not using toxic chemicals – you will be helping other life forms to thrive in your growing area. Leaving some weeds, such as a discreet area of stinging nettles, will provide food for pollinating insects, as well as leaves to make a liquid feed, high in nitrates.

Direct Weed Control

Although cultural methods provide the basis for weed management in organic crop rotations it is likely that some form of direct action will be needed against weeds to prevent crop loss at some time. Before taking any action it is important to take an overview and assess whether the weeds present are likely to develop to such an extent that they will cause an immediate loss of crop or will store up potential future problems (e.g. by shedding seed and adding to the soil seed bank so exacerbating future weed problems). If the weed burden is judged to have the potential to cause damage the cost of this should be offset against the likely costs of any immediate or future direct control measures so that direct weeding is only underaken when it is economically beneficial to do so.

- Mechanical weed control provides an overview of the range of options and implements available for direct mechanical weed control in the field.

- Manual weed control is still an important component of many weed management programmes and this topic provides an overview of techniques available for use on farm.

- Thermal weed control is becoming more popular and is described in this section.

- Mulching provides a physical barrier to weed development and is often used in horticultural crops to control weeds.

- Biological weed control aims to get insects, pathogens or even other plants to do the work of weed management for the farmer.

- Allelopathy can be regarded as a component of biological control in which plants are used to reduce the vigour and development of other plants.

Mechanical Weed Control

Mechanical weed control may involve weeding the whole crop, or it may be limited to selective inter-row or intra-row weeding. Machines can be used to kill weeds by burying, cutting or uprooting. Tools without a cutting action are only effective on small weeds. Inter-row implements have been designed that control weeds within the crop row by directing soil along the row to cover small weeds. With slow germinating crops inter-row weeding may have to be delayed until the crop seedlings emerge. In some situations it may be possible to include a few seeds of a fast germinating crop in the seed mixture to give an early indication of the position of the crop row.

Mechanical weeders range from basic hand tools to sophisticated tractor driven or self-propelled devices. These may include cultivating tools such as hoes, harrows, tines and brush weeders, cutting tools like mowers and strimmers, as well as implements like thistle-bars that may do both. Two wheeled pedestrian or walking tractors are a smaller alternative that can power a similar range of implements. Custom-made basket or cage-wheeled weeders, with gangs of rolling wire cylinders, offer another way to deal with seedling weeds in a friable soil. The choice of implement, and the timing and frequency of its use may depend on the crop and on the weed population. Some implements, such as fixed harrows, are thought more suitable for arable crops, while others like inter-row brush weeders may be considered to be more effective for horticultural use.

The weather and soil conditions under which the operation is carried out will have a major influence on its efficacy. Soil type, surface structure and moisture content affect the choice and efficacy of mechanical weed control implements. The options may be more limited on heavy or stony soils, most implements work better on light, stone-free soils. Mechanical weeding is less effective when soils are wet during or after weeding operations. Some implements like brushweeders are able to work at a higher soil moisture content than others. Conversely, brushweeders do not work well on dry soils where the surface has capped because the brushes cannot penetrate the crust.

The optimum timing for mechanical weed control is influenced by the competitive ability of the crop. A single inter-row cultivation at any time may provide excellent weed control in a crop like transplanted broccoli that rapidly develops a broad, shading leaf canopy but may be poorer in crops like sweet corn (Zea mays) where early growth is slow, or in green beans (Phaseolus vulgaris) where the growing season is relatively long. In the UK, the optimum timings for mechanical weed control have been defined for onions and for carrots grown in both organic and conventional systems. In organic winter cereals studies have found that , found that corn poppy (Papaver rhoeas) was more effectively controlled in the autumn whilst chickweed (S. media) was controlled best in spring when using a spring-tine weeder.

Weed morphology and stage of growth will also influence the selection and efficacy of weeding implement. In experiments to determine the type of physical damage that gave the most effective control of a range of seedling weeds it has been found that burial to 1 cm depth is the most effective treatment, closely followed by cutting at the soil surface. Plants need to be buried totally to be killed but plant size, angle and growth habit influence the depth of covering required. Some advisors suggest that if weeds have emerged you are already late with your weeding operation, and that the best time to kill them is at the white thread stage. At high weed densities, even with the most effective mechanical weeders, sufficient weeds are likely to survive control measures and profoundly reduce crop yield in cereals and direct control needs to be linked with long term preventative measures to maintain the weed population at a manageable level.

With most mechanical weeding implements, operator skill, experience and knowledge are critical to success. Drawbacks to mechanical weed control include low work rates, delays due to wet conditions, and the subsequent risk of weed control failure as weeds become larger. A review of the merits of six different mechanical weeding mechanisms in controlling inter-row weeds at different growth stages and at different tractor speeds indicated that weed control was not necessarily better at earlier weed stages and weeding too early often missed late germinating weeds. Increasing forward speed did not improve the performance of all the implements equally.

There may be some disadvantages to the greater use of mechanical weed control. The additional cultivations associated with mechanical weeding could harm soil structure and possibly encourage soil erosion. The increased mineralisation of soil nitrogen due to cultivation may be seen by some growers as a problem and by others as an advantage (although this is likely to be limited in effect). There is concern about the impact of mechanical weeding on ground nesting birds and management practices may some alteration to minimise disruption at critical times although evidence is at times contradictory. Any soil cultivation will also contribute to the movement of weed seeds in the soil. Studies of the horizontal movement of freshly shed seeds have shown that a sequence of cultivations could move seeds over 2 m horizontally.

Manual Weed Control

Manual methods of weed control are still widely used in organic systems. Hand weeding is most useful on annual weeds and some perennial weeds. There are times when hand roguing individual plants or patches of plants is the most effective way of preventing them spreading and multiplying. It is widely used for dealing with the removal of difficult-to-control species such as docks, thistles and ragwort.

Manual methods of weed control are also widely used in intensive horticultural crops where it is important to perform a good first weed to prevent weed competition. Hand weeding can often follow after a mechanical inter-row weeding operation in order to thoroughly remove weeds in the crop row. It is a practical method of removing weeds within rows and hills where a cultivating implement cannot be used. It obviously requires more labour than other direct weed control methods and therefore costs are likely to be higher so it is only employed by growers with high value crops like vegetables.

It is generally more efficient for groups or gangs of workers to hoe or hand weed crops as a team, whether directly pulling the weeds or using some type of hoe. Hand rogueing or pulling is a widely used technique for patches of weeds or removing. There are many modern hoe designs that are more comfortable to use than traditional designs and these should be investigated where large areas are being covered. Some designs include the stirrup hoe, the diamond hoe and the collinear hoe as well as wheeled push hoes. Other tools have been designed to tackle specific weeds such as docks, thistles or ragwort.

In more mechanised systems teams of workers lie on a flat bed weeder pulled by a tractor or on other specially designed machines. The speed of the machines can be adjusted to accommodate the level of weeds in the crop.

Thermal Weed Control

One of the earliest forms of thermal weed control, stubble burning, is now banned because of the smoke and other hazards it created. However, this traditional form of thermal weed control was effective in reducing the number of viable weed seeds returned to the soil after cereal harvest. Soil surface temperatures under the burning straw reached in excess of 200 °C for 10-30 seconds and reduced the viability of freshly shed wild oat (Avena fatua) and blackgrass (Alopecurus myosuroides) seed by up to 30% and 80% respectively. Current thermal weed control methods use a variety of thermal weeders to generate the heat needed to kill weed seeds and weed seedlings.

Flaming equipment has been developed in several countries including Germany, Holland, Sweden and Denmark, and a range of tractor and smaller hand operated burners is available in the UK. The main fuel used in the burners is liquefied petroleum gas (LPG) usually propane.

Mulching

Covering or mulching the soil surface can reduce weed problems by preventing weed seed germination or by suppressing the growth of emerging seedlings. Mulches are generally ineffective against established perennial weeds. A mulch may take many forms: a living plant ground cover, loose particles of organic or inorganic matter spread over soil, and sheets of artificial or natural materials laid on the soil surface. Residues from preceding crops may be used to form a mulch but this is discussed in more detail in the use of cover crops to suppress weeds. With mulches consisting of organic materials, crop stand and vigour, particularly of direct-seeded small-seeded crops, may be reduced by chemicals released from the decomposing residues.

It is most practical to use mulches in well-spaced crops, particularly transplants. Plastic sheeting and straw mulches have long been used in soft fruit such as strawberries. In perennial crops and some other situations mulches may be intended to remain effective for many years. Mypex, a black, woven, polypropylene mulch, is expected to last for up to three crops (9-10 years). These mulches may be expensive but labour costs are reduced in the long term. Other uses for mulches include as an alternative to cultivation to clear vegetation before cropping by leaving them in place for 12 to 18 months. In freshly prepared seedbeds, short term mulching can be used to manipulate or reduce weed seedling emergence, by for instance, laying black plastic on the seedbed 2 to 8 weeks and then lifting it before planting brassicas or other crops.

The high cost of mulching makes it economic only for high value horticultural crops unless there is another reason for its use. In addition to weed control, mulches may be used: to prevent soil erosion, reduce pest problems, to aid moisture and to prevent nitrate loss. In strawberries, rain splash dispersal of disease spores like those of black spot (Colletotrichum acutatum) is reduced by straw mulch. Mulches can also moderate soil temperatures. Organic mulches in particular reduce heat loss from the soil in cold conditions and help to prevent frost heave. In hot weather the mulch slows down the warming of soil.

Types of Mulches

Living mulches: consist of a dense stand of low growing plants established prior to or after the crop. The undersowing of cereals with clover and grass could be seen as forming a living mulch. It has been argued that annual weeds would provide a natural ground cover if managed properly. Living mulches are sometimes referred to as cover crops, but they grow at least part of the time simultaneously with the crop. Cover crops are generally killed off prior to crop establishment.

Often, the primary purpose of a living mulch is that of improving soil structure, aiding nutrition or avoiding pest attack, and weed suppression may be just an added benefit. Maintaining vegetation cover is important for preventing soil erosion, nitrate leaching and weed emergence in slowly developing crops like maize. An investigation of the influence of different mulch species on weed density and diversity indicated that weed numbers were reduced and maize yield was not affected where growth of the mulch was reduced by cutting or flaming treatments. When the growth of a living mulch is not restricted, or when soil moisture is inadequate, even a relatively vigorous crop like potato may suffer competition and loss of yield. Studies have been made of the use of living mulches to suppress weed emergence in horticultural crops but there are many different factors to take into account and it is difficult to get the balance between crop and mulch right. Living mulches are well suited to use in perennial crops such as fruit where self-reseeding is an advantage. However, even in established apple and apricot orchards a living mulch growing along the planted row may depress crop growth. In amenity situations, ground covering plants are established to form a dense canopy and suppress weed germination and growth.

Particle mulches: may be organic or inorganic. Loose materials like straw, bark and composted municipal green waste provide effective weed control but the depth of mulch needed to suppress weed emergence is likely to make transport costs prohibitive unless the material is produced on the farm. It has been shown that a 3 cm layer of compost was needed to prevent the emergence of annual weeds and weed control usually improves as the thickness of the organic mulch increases. Weed seeds in the mulch itself can be a problem if the composting process has not been fully effective or there is contamination by wind blown seeds. In straw mulches, volunteer cereal seedlings are a particular problem due to shed cereal grains and even whole ears remaining in the straw after crop harvest. With particle mulches like straw that consist of light materials there is the possibility of them being blown around by the wind. Organic mulches like straw with a high carbon to nitrogen ratio may deplete the soil of nitrogen as they decompose. Mulch improves water filtration into the soil and prevents the compaction and erosion that heavy rainfall can cause.

Before applying a particle mulch weeds should be removed, dry soil should be moistened, and compacted soil loosened. Old mulch should be removed or incorporated to prevent a build-up. Most mulch is applied 7.5 to 10 cm deep. Coarser textured materials require thicker layers. On sandy soils, the mulch layer needs to be deeper than on heavy or wet soils. The mulch should be raked periodically and topped up if necessary. Machinery has been developed for applying/spreading particle mulches. Bark blowers are widely available. Self-feeding straw blowers can handle 1-2 bales per minute and can cover an acre per hour. Flail-beater chains break the straw into 5-10 cm lengths. For green waste there are all-in-one collectors, shredders and spreaders.

Sheeted mulches: a layer of material such as plastic, paper of woven fabric covers the soil surface. Black polyethylene mulches are widely used for weed control in organic and conventional systems in the UK and elsewhere. Clear mulches are better than black for warming the soil but do not control the weeds. Plastic mulches have been developed that selectively filter out the photosynthetically active radiation (PAR) but let through infra red light to warm the soil. Infra red transmitting (IRT) mulches have been shown to be effective in controlling weeds. Various colours of woven and solid film plastics have been tested in the field. White and green coverings had little effect on the weeds, brown, black, blue, and white on black (double colour) films prevented weeds emerging. There are indications that mulching films, like white on black, with a higher rate of light reflectance are beneficial to the crop. Light reflectance may also affect the behaviour of certain insects, and plastic mulches in a greater array of colours are likely to become available. The woven and non-woven polypropylene films or geotextiles (like Mypex) are sometimes referred to as weed barriers and landscape fabrics. They are more durable than polyethylene films permitting multi-year use and are permeable to water. There are advantages both in reduced laying and disposal costs compared with single season materials.

Sheeted materials are relatively expensive and are usually laid by machine. Machinery has been developed that will raise the soil into beds and lay the plastic mulch, securing it at the edges. Beds can be prepared in advance of crop planting. Heavy duty plastic is used for long term crops such as perennial herbs. Woven polypropylene fabrics allow water to penetrate and are less likely to scorch crops when temperatures are high. Non-woven black fabric mulch may not be sufficiently opaque to prevent weed growth completely. After cropping, lifting and disposal may be a problem with plastic and other durable mulches and this adds to the overall costs. Even the degradable plastics may break into fragments that litter the soil. Sheeting made from paper and other natural fibres have the advantage of breaking down naturally, and can be incorporated into the soil after use. Paper mulches have compared favourably with black polyethylene in trials with transplanted lettuce, Chinese cabbage and calabrese in the UK although tearing and wind blowing can be a problem.

Biodegradable film mulches have been developed that adjust their degradation according to the soil and weather conditions. The polymers are strong enough to withstand mechanical laying.

Biological Weed Control

Biological weed control involves the release of organisms that attack plants to control weeds. The aim of biological control is to shift the balance of competition between the weed and the crop in favour of the crop and against the weed. The biological control agent, normally a fungus or insect, may not necessarily kill the target weed but should, at the least, reduce its vigour and competitive ability. From a practical point of view the organism or agent should prevent the weed setting seed or producing other reproductive parts. There is considerable potential for encouraging the use of native biological control agents against weeds and substantial research effort has been put into biological control in general. However, the application of biological weed control in agricultural systems in Europe has proved difficult and their are no well documented successes.

In practice, there are three basic types of biological control:

Classical (or innoculative) biological control involves the release of exotic natural enemies to control exotic weeds and has been successful against weeds like thistles in the US and Australia where weevils (native to Europe) have been introduced onto the thistles. It has been suggested that some introduced weeds like hogweed (Heracleum mantegazzium), Himalayan balsam (Impatiens glandulifera) and the Japanese knotweeds (Reynoutria spp.) would be ideal candidates for classical biological control but so far it has only been attempted with bracken (Pteridium aquilinum) where attempts to use two South African moths as potential biological control agents were not successful.

The introduction of a classical biological control agent may not be deliberate. A rust, Puccinia lagenophorae, of Australian origin, which attacks a range of Senecio species, was unknown in Europe before 1960 but has since become established in France and the UK on groundsel where it reduces the viability of groundsel plants on which it can be regularly found. It is unknown how this pathogen reached Europe or how it established.

Inundative control involves the mass production and release of native natural enemies against native weeds, for example rust fungus is often used against weeds. Work in this area has concentrated on fungal pathogens of plants as they can potentially be applied as sprays in the same way as conventional herbicides (hence their name myco- or bio-herbicides). Studies on bioherbicides have concentrated mainly on foliar treatments using fungi. Commercial products have been developed (mainly in the US) but success has been limited Soil micro-organisms are often overlooked but are also important as plant pathogens. Several are being investigated as potential biological control agents particularly for control of grass weeds such as downy brome, wild oat and green foxtail.

Although much of the work on biological control agents has concentrated on the growing weed plant, there is considerable potential for using micro-organisms to manipulate or deplete the soil weed seedbank. The persistence of weed seeds in the soil is the key to their success in continuing to emerge despite repeated control measures over many years. Greater predation or an increase in natural decay would reduce the soil seedbank and hence future weed populations. However, there are as of yet no practical or commercial applications available.

Conservation control is an indirect method, which manipulates the habitat around the weeds with the aim of encourging those organisms that attack the weed. This is a long term strategy that requires a detailed knowledge of the ecology of the crop weed habitat, the target weeds and the control agents. It has received little attention to date. One recent example is the upsurge of interest in looking at encouraging the dock beetle on dock plants by creating conditions that favour the beetle.

Livestock can also be considered as biological control agents which can give a broad spectrum control of weeds in various situations.

The assessment of the potential risks involved in introducing biological control agents remains a difficult and (sometimes) contentious issue as any predictions of how biological control may affect the interaction between species, and influence the life cycle of non-target species is extremely complicated. Even if there were no risk to non-target species, there could still be a conflict of interests because some may perceive a particular plant as a weed while others see it as a desirable wild flower, or even a potential crop. For this reason it seems difficult to imagine that of the shelf biological control of weeds is a realistic prospect in the short to medium term.

Allelopathy

Allelopathy can be regarded as a component of biological control in which plants are used to reduce the vigour and development of other plants. Allelopathy refers to the direct or indirect chemical effects of one plant on the germination, growth, or development of neighbouring plants. This can be through the release of allelochemicals while the plant is growing or from plant residues as it rots down. These chemicals can be released from around the germinating seed, in exudates from plant roots, from leachates in the aerial part of the plant and in volatile emissions from the growing plant. Both crops and weeds are capable of producing these compounds and in this case the desired effect is the impaired germination, reduced growth and poor development of weeds.

Potentially allelopathy could be used in various ways:

- To manipulate the crop-weed balance by increasing the toxicity of the crop plants to weeds thereby reducing weed germination in the direct area of the crop, which is the most difficult area to control physically.

- As cover crops to suppress weed germination and development over a whole field in part of a rotation.

- As mulched residues or incorporated residues which could prevent weed germination and allow transplanted crops to be grown, producing a residual weed control effect.

Many crops have been reported as showing allelopathic properties at one time or another and farmers report that some crops such as oats seem to clean fields of weeds better than others. The current list includes: wheat, barley, oats, cereal rye, brassicas, red clover, yellow sweet clover, trefoil, vetch, buckwheat, lucerne, rice, sorghum.

However, several weed species have also been reported to show allelopathic properties. They include couch grass, creeping thistle and chickweed. Where they occur together they may have a synergistic negative effect on crops.

Allelopathic effects might also depend on a number of other factors that might be important in any given situation:

- Varieties: There can be a great deal of difference in the strength of allelopathic effects between different crop varieties.

- Specificity: There is a significant degree of specificity in allelopathic effects. Thus, a crop which is strongly allelopathic against one weed may show little or no effect against another.

- Autotoxicity: Allelopathic chemicals may not only suppress the growth of other plant species, they can also suppress the germination or growth of seeds and plants of the same species. Lucerne is particularly well known for this and has been well researched. The toxic effect of wheat straw on following wheat crops is also well known.

- Crop on crop effects: Residues from allelopathic crops can hinder germination and growth of following crops as well as weeds. A sufficient gap must be left before the following crop is sown. Larger seeded crops are effected less and transplants are not affected.

- Environmental factors: Several factors impact on the strength of the allelopathic effect. These include pests and disease and especially soil fertility. Low fertility increases the production of allelochemicals. After incorporation the alleopathic effect declines fastest in warm wet conditions and slowest in cold wet conditions.

Mulch

A mulch is a layer of material applied to the surface of soil. Reasons for applying mulch include conservation of soil moisture, improving fertility and health of the soil, reducing weed growth and enhancing the visual appeal of the area.

A mulch is usually, but not exclusively, organic in nature. It may be permanent (e.g. plastic sheeting) or temporary (e.g. bark chips). It may be applied to bare soil or around existing plants. Mulches of manure or compost will be incorporated naturally into the soil by the activity of worms and other organisms. The process is used both in commercial crop production and in gardening, and when applied correctly, can dramatically improve soil productivity.

Uses

Many materials are used as mulches, which are used to retain soil moisture, regulate soil temperature, suppress weed growth, and for aesthetics. They are applied to the soil surface, around trees, paths, flower beds, to prevent soil erosion on slopes, and in production areas for flower and vegetable crops. Mulch layers are normally 2 inches (5.1 cm) or more deep when applied.

They are applied at various times of the year depending on the purpose. Towards the beginning of the growing season, mulches serve initially to warm the soil by helping it retain heat which is lost during the night. This allows early seeding and transplanting of certain crops, and encourages faster growth. As the season progresses, mulch stabilizes the soil temperature and moisture, and prevents the growing of weeds from seeds. In temperate climates, the effect of mulch is dependent upon the time of year they are applied and when applied in fall and winter, are used to delay the growth of perennial plants in the spring or prevent growth in winter during warm spells, which limits freeze thaw damage.

The effect of mulch upon soil moisture content is complex. Mulch forms a layer between the soil and the atmosphere preventing sunlight from reaching the soil surface, thus reducing evaporation. However, mulch can also prevent water from reaching the soil by absorbing or blocking water from light rains.

In order to maximise the benefits of mulch, while minimizing its negative influences, it is often applied in late spring/early summer when soil temperatures have risen sufficiently, but soil moisture content is still relatively high. However, permanent mulch is also widely used and valued for its simplicity, as popularized by author Ruth Stout, who said, "My way is simply to keep a thick mulch of any vegetable matter that rots on both sides of my vegetable and flower garden all year long. As it decays and enriches the soils, I add more."

Plastic mulch used in large-scale commercial production is laid down with a tractor-drawn or standalone layer of plastic mulch. This is usually part of a sophisticated mechanical process, where raised beds are formed, plastic is rolled out on top, and seedlings are transplanted through it. Drip irrigation is often required, with drip tape laid under the plastic, as plastic mulch is impermeable to water.

Materials

Rubber mulch nuggets in a playground. The white fibers are nylon cords, which are present in the tires from which the mulch is made.

Shredded wood used as mulch. This type of mulch is
often dyed to improve its appearance in the landscape.

Materials used as mulches vary and depend on a number of factors. Use takes into consideration availability, cost, appearance, the effect it has on the soil—including chemical reactions and pH, durability, combustibility, rate of decomposition, how clean it is—some can contain weed seeds or plant pathogens.

Pine needles used as mulch. Also called pinestraw.

A variety of materials are used as mulch:

- Organic residues: Grass clippings, leaves, hay, straw, kitchen scraps comfrey, shredded bark, whole bark nuggets, sawdust, shells, woodchips, shredded newspaper, cardboard, wool, animal manure, etc. Many of these materials also act as a direct composting system, such as the mulched clippings of a mulching lawn mower, or other organics applied as sheet composting.

- Compost: Fully composted materials are used to avoid possible phytotoxicity problems. Materials that are free of seeds are ideally used, to prevent weeds being introduced by the mulch.

- Old carpet (synthetic or natural): Makes a free, readily available mulch.

- Rubber mulch: Made from recycled tire rubber.

- Plastic mulch: Crops grow through slits or holes in thin plastic sheeting. This method is predominant in large-scale vegetable growing, with millions of acres cultivated under plastic mulch worldwide each year (disposal of plastic mulch is cited as an environmental problem).

- Rock and gravel can also be used as a mulch. In cooler climates the heat retained by rocks may extend the growing season.

Spring daffodils push through shredded wood mulch.

In some areas of the United States, such as central Pennsylvania and northern California, mulch is often referred to as "tanbark", even by manufacturers and distributors. In these areas, the word "mulch" is used specifically to refer to very fine tanbark or peat moss.

Organic Mulches

Mulching coconut farm.

Organic mulches decay over time and are temporary. The way a particular organic mulch decomposes and reacts to wetting by rain and dew affects its usefulness.

Some mulches such as straw, peat, sawdust and other wood products may for a while negatively affect plant growth because of their wide carbon to nitrogen ratio, because bacteria and fungi that decompose the materials remove nitrogen from the surrounding soil for growth. However, whether this effect has any practical impact on gardens is disputed by researchers and the experience of gardeners. Organic mulches can mat down, forming a barrier that blocks water and air flow between the soil and the atmosphere. Vertically applied organic mulches can wick water from the soil to the surface, which can dry out the soil. Mulch made with wood can contain or feed termites, so care must be taken about not placing mulch too close to houses or building that can be damaged by those insects. Some mulch manufacturers recommend putting mulch several inches away from buildings.

Commonly available organic mulches include:

Leaves

Leaves from deciduous trees, which drop their foliage in the autumn/fall. They tend to be dry and blow around in the wind, so are often chopped or shredded before application. As they decompose they adhere to each other but also allow water and moisture to seep down to the soil surface. Thick layers of entire leaves, especially of maples and oaks, can form a soggy mat in winter and spring which can impede the new growth lawn grass and other plants. Dry leaves are used as winter mulches to protect plants from freezing and thawing in areas with cold winters; they are normally removed during spring.

Grass Clippings

Grass clippings, from mowed lawns are sometimes collected and used elsewhere as mulch. Grass clippings are dense and tend to mat down, so are mixed with tree leaves or rough compost to provide aeration and to facilitate their decomposition without smelly putrefaction. Rotting fresh grass clippings can damage plants; their rotting often produces a damaging buildup of trapped heat. Grass clippings are often dried thoroughly before application, which mediates against rapid decomposition and excessive heat generation. Fresh green grass clippings are relatively high in nitrate content, and when used as a mulch, much of the nitrate is returned to the soil, conversely the routine removal of grass clippings from the lawn results in nitrogen deficiency for the lawn.

Peat Moss

Peat moss, or sphagnum peat, is long lasting and packaged, making it convenient and popular as a mulch. When wetted and dried, it can form a dense crust that does not allow water to soak in. When dry it can also burn, producing a smoldering fire. It is sometimes mixed with pine needles to produce a mulch that is friable. It can also lower the pH of the soil surface, making it useful as a mulch under acid loving plants.

However peat bogs are a valuable wildlife habitat, and peat is also one of the largest stores of carbon (in Britain, out of a total estimated 9952 million tonnes of carbon in British vegetation and soils, 6948 million tonnes carbon are estimated to be in Scottish, mostly peatland, soils), so gardeners who wish to protect the environment will choose more sustainable alternatives.

Wood Chips

Wood chips are a byproduct of the pruning of trees by arborists, utilities and parks; they are used to dispose of bulky waste. Tree branches and large stems are rather coarse after chipping and tend to be used as a mulch at least three inches thick. The chips are used to conserve soil moisture, moderate soil temperature and suppress weed growth. The decay of freshly produced chips from recently living woody plants, consumes nitrate; this is often off set with a light application of a high-nitrate fertilizer. Wood chips are most often used under trees and shrubs. When used around soft stemmed plants, an unmulched zone is left around the plant stems to prevent stem rot or other possible diseases. They are often used to mulch trails, because they are readily produced with little additional cost outside of the normal disposal cost of tree maintenance. Wood chips come in various colors.

Woodchip Mulch

Woodchip mulch is a byproduct of reprocessing used (untreated) timber (usually packaging pallets), to dispose of wood waste by creating woodchip mulch. The chips are used to conserve soil moisture, moderate soil temperature and suppress weed growth. Woodchip mulch is often used under trees, shrubs or large planting areas and can last much longer than arborist mulch. In addition, many consider wood-chip mulch to be visually appealing, as it comes in various colors. Woodchips can also be reprocessed into playground woodchip to be used as an impact-attenuating playground surfacing.

Bark Chips

Bark chips.

Bark chips of various grades are produced from the outer corky bark layer of timber trees. Sizes vary from thin shredded strands to large coarse blocks. The finer types are very attractive but have a large exposed surface area that leads to quicker decay. Layers two or three inches deep are usually used, bark is relativity inert and its decay does not demand soil nitrates. Bark chips are also available in various colors.

Straw Mulch

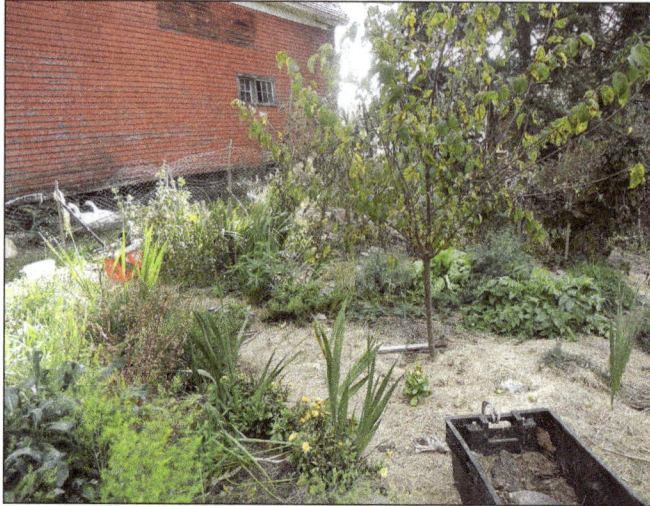

Permaculture garden with a fruit tree, herbs, flowers and
vegetables mulched with hay.

Straw mulch or *field hay* or *salt hay* are lightweight and normally sold in compressed
bales. They have an unkempt look and are used in vegetable gardens and as a winter
covering. They are biodegradable and neutral in pH. They have good moisture retention and weed controlling properties but also are more likely to be contaminated with
weed seeds. Salt hay is less likely to have weed seeds than field hay. Straw mulch is also
available in various colors.

Cardboard

Cardboard or *newspaper* can be used as mulches. These are best used as a base layer
upon which a heavier mulch such as compost is placed to prevent the lighter cardboard/
newspaper layer from blowing away. By incorporating a layer of cardboard/newspaper
into a mulch, the quantity of heavier mulch can be reduced, whilst improving the weed
suppressant and moisture retaining properties of the mulch. However, additional labour is expended when planting through a mulch containing a cardboard/newspaper
layer, as holes must be cut for each plant. Sowing seed through mulches containing a
cardboard/newspaper layer is impractical. Application of newspaper mulch in windy
weather can be facilitated by briefly pre-soaking the newspaper in water to increase its
weight.

Carpet

Synthetic carpet that is composed of artificial fibers may be removed after planting to
prevent fibers taking a long time to decompose, whereas carpet made from natural fibers may be kept in place, blocking competition from weeds. Rain is absorbed by carpet
and then slowly released into the soil, reducing watering needs.

Colored Mulch

Some organic mulches are colored red, brown, black, and other colors. Isopropanol-amine, specifically 1-Amino-2-propanol or DOW monoisopropanolamine, may be used as a pigment dispersant and color fastener in these mulches. Types of mulch which can be dyed include: wood chips, bark chips (barkdust) and pine straw. Colored mulch is made by dyeing the mulch in a water-based solution of colorant and chemical binder. When colored mulch first entered the market, most formulas were suspected to contain toxic, heavy metals and other contaminates. Today, "current investigations indicate that mulch colorants pose no threat to people, pets or the environment. The dyes currently used by the mulch and soil industry are similar to those used in the cosmetic and other manufacturing industries (i.e., iron oxide)," as stated by the Mulch and Soil Council. Colored mulch can be applied anywhere non-colored mulch is used (such as large bedded areas or around plants) and features many of the same gardening benefits as traditional mulch, such as improving soil productivity and retaining moisture. As mulch decomposes, just as with non-colored mulch, more mulch may need to be added to continue providing benefits to the soil and plants. However, if mulch is faded, spraying dye to previously spread mulch in order to restore color is an option.

Anaerobic Mulch

Mulch normally smells like freshly cut wood, but sometimes develops a toxicity that causes it to smell like vinegar, ammonia, sulfur or silage. This happens when material with ample nitrogen content is not rotated often enough and it forms pockets of increased decomposition. When this occurs, the process may become anaerobic and produce these phytotoxic materials in small quantities. Once exposed to the air, the process quickly reverts to an aerobic process, but these toxic materials may be present for a period of time. If the mulch is placed around plants before the toxicity has had a chance to dissipate, then the plants could very likely be damaged or killed depending on their hardiness. Plants that are predominantly low to the ground or freshly planted are the most susceptible, and the phytotoxicity may prevent germination of some seeds.

If sour mulch is applied and there is plant kill, the best thing to do is to water the mulch heavily. Water dissipates the chemicals faster and refreshes the plants. Removing the offending mulch may have little effect, because by the time plant kill is noticed, most of the toxicity is already dissipated. While testing after plant kill will not likely turn up anything, a simple pH check may reveal high acidity, in the range of 3.8 to 5.6 instead of the normal range of 6.0 to 7.2. Finally, placing a bit of the offending mulch around another plant to check for plant kill will verify if the toxicity has departed. If the new plant is also killed, then sour mulch is probably not the problem.

Groundcovers

Groundcovers are plants which grow close to the ground, under the main crop, to slow the development of weeds and provide other benefits of mulch. They are usually fast-growing plants that continue growing with the main crops. By contrast, cover crops are incorporated into the soil or killed with herbicides. However, live mulches also may need to be mechanically or chemically killed eventually to prevent competition with the main crop.

Some groundcovers can perform additional roles in the garden such as nitrogen fixation in the case of clovers, dynamic accumulation of nutrients from the subsoil in the case of creeping comfrey (*Symphytum ibericum*), and even food production in the case of *Rubus tricolor*.

On-site Production

Owing to the great bulk of mulch which is often required on a site, it is often impractical and expensive to source and import sufficient mulch materials. An alternative to importing mulch materials is to grow them on site in a "mulch garden" – an area of the site dedicated entirely to the production of mulch which is then transferred to the growing area. Mulch gardens should be sited as close as possible to the growing area so as to facilitate transfer of mulch materials.

Mulching over Unwanted Plants

Sufficient mulch over plants will destroy them, and may be more advantageous than using herbicide, cutting, mowing, pulling, raking, or tilling. The higher the temperature that this "mulch" is composted, the quicker the reduction of undesirable materials. "Undesirable materials" may include living seed, plant "trash", as well as pathogens such as from animal feces, urine (e.g. hantavirus), fleas, lice, ticks, etc.

In some ways this improves the soil by attracting and feeding earthworms, and adding humus. Earthworms "till" the soil, and their feces are among the best fertilizers and soil conditioners. Urine may be toxic to plants if applied to growing areas undiluted.

Polypropylene and Polyethylene Mulch

Polypropylene mulch is made up of polypropylene polymers where polyethylene mulch is made up of polyethylene polymers. These mulches are commonly used in many plastics. Polyethylene is used mainly for weed reduction, where polypropylene is used mainly on perennials. This mulch is placed on top of the soil and can be done by machine or hand with pegs to keep the mulch tight against the soil. This mulch can prevent soil erosion, reduce weeding, conserve soil moisture, and increase temperature of the soil. Ultimately this can reduce the amount of work a farmer may have to do, and the amount of herbicides applied during the growing period. The black and clear mulches capture sunlight and warm the soil increasing the growth rate. White and other reflective colours will also warm the soil, but they do not suppress weeds as well. This mulch may require other sources of obtaining water such as drip irrigation since it can reduce the amount of water that reaches the soil. This mulch needs to be manually removed at the end of the season since when it starts to break down it breaks down into smaller pieces. If the mulch is not removed before it starts to break down eventually it will break down into ketones and aldehydes polluting the soil. This mulch is technically biodegradable but does not break down into the same materials the more natural biodegradable mulch does.

Biodegradable Mulch

Quality biodegradable mulches are made out of plant starches and sugars or polyester fibers. These starches can come from plants such as wheat and corn. These mulch films may be a bit more permeable allowing more water into the soil. This mulch can prevent soil erosion, reduce weeding, conserve soil moisture, and increase temperature of the soil. Ultimately this can reduce the amount of herbicides used and manual labor farmers may have to do throughout the growing season. At the end of the season these mulches will start to break down from heat. Microorganisms in the soil break down the mulch into two components, water and CO_2, leaving no toxic residues behind. This source of mulch is even less manual labor since it does not need to be removed at the end of the season and can actually be tilled into the soil. With this mulch it is important to take into consideration that it's much more delicate than other kinds. It should be placed on a day which is not too hot and with less tension than other synthetic mulches. These also can be placed by machine or hand but it is ideal to have a more starchy mulch that will allow it to stick to the soil better.

Sheet Mulching

Sheet mulching is an agricultural no-dig gardening technique that attempts to mimic the natural soil-building process in forests. When deployed properly and in combination with other permaculture principles, it can generate healthy, productive, and low maintenance ecosystems.

Sheet mulching, also known as composting in place, mimics nature by breaking down organic material from the topmost layers down. The simplest form of sheet mulching consists of applying a bottom layer of decomposable material, such as cardboard or newspapers, to the ground to kill existing vegetation and suppress weeds. Then, a top layer of organic mulch is applied. More elaborate sheet mulching involves more layers. Sheet mulching is used to transform a variety of surfaces into a fertile soil that can be planted. Sheet mulching can be applied to a lawn, a dirt lot full of perennial weeds, an area with poor soil, or even pavement or a rooftop.

Technique

A model for sheet mulching consists of the following steps:

- The area of interest is flattened by trimming down existing plant species such as grasses.

- The soil is analyzed and its pH is adjusted (if needed).

- The soil is moisturized (if needed) to facilitate the activity of decomposers.

- The soil is then covered with a thin layer of slowly decomposing material (known as the weed barrier), typically cardboard. This suppresses the weeds by blocking sunlight, adds nutrients to the soil as weed matter quickly decays beneath the barrier, and increases the mechanical stability of the growing medium.

- A layer (around 10 cm thick) of weed-free soil, rich in nutrients is added, in an attempt to mimic the surface soil, or A horizon.

- A layer (at most 15 cm thick) of weed-free, woody and leafy matter is added in an attempt to mimic the forest floor, or O horizon. Theoretically, the soil is now ready to receive the desirable plant seeds or transplants.

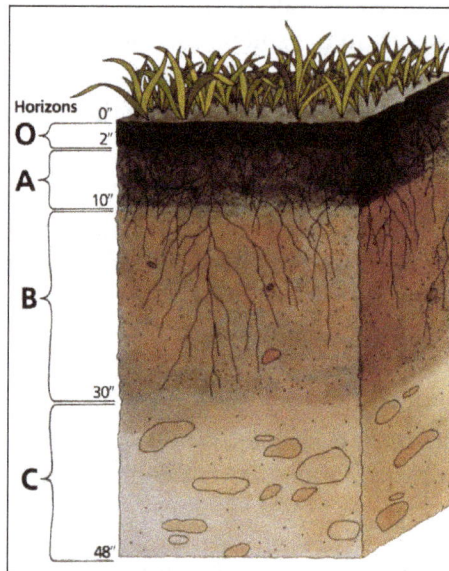

Typical layers of natural soil.

Variations and Considerations

- Often the barrier is applied a few months before planting to ensure the penetration of roots of newly planted seeds.

- Very thick barriers can cause anaerobic conditions.

- Some permaculturists incorporate composting in steps 5 and/or 6.

- Sheets of newspaper and clothing can be used instead of cardboard.

- An initial layer (2–3 kg/m²) of matter rich in nutrients (such as compost or manure) may be added to bolster decomposition.

- Some varieties of grasses and weeds may be beneficial in a number of ways. Such plants can be controlled and used rather than eradicated. One variation of mulching, called hugelkultur, involves using buried logs and branches as the first layer of the bed.

Advantages

Sheet mulch has important advantages relative to conventional methods, such as tilling, plowing or applying herbicides:

- Improvement of desirable plants' health and productivity.

- Retention of water and nutrients and stabilization of biochemical cycles.

- Improvement of soil structure, soil life, and prevention of soil erosion.

- Avoidance of potentially dangerous pesticides, especially herbicides.

- Reduction of overall maintenance labor and costs.

- Most of the materials required to sheet mulch can be collected at no cost, and materials can be substituted for those readily available in certain areas. For instance, suburban areas may have a plentiful supply of leaves, and farming communities may have spoiled hay and manure.

Disadvantages

- Some weed seeds (such as those of Bermuda grass and species of bindweed) may persist under the barrier and within the soil seed bank.

- Termites are attracted to the area. While they are a natural part of the ecosystem that transforms the weed barrier into rich soil, they can pose a hazard to nearby wood-framed structures.

- Slug populations may increase during the early stages of decomposition. However they can be kept away or harvested.

- The system may need a constant supply of organic material, at least during the early stages.

- Roaming animals may interrupt the sheet mulching process.

Cultural Weed Control

Organic farmers recognise that every element of farming is inter-linked, and that good rotational design produces healthy soil, healthy plants and good yields. Crop rotation is the cornerstone of organic farming practice. Rotation and forward planning are also important for managing weeds. In this topic we provide information on cultural methods aimed at preventing weed problems arising in the first place and which farmers and growers can plan to incorporate into their rotations.

The underlying principle of a preventative approach is to produce a constantly changing environment to which no single weed species can adapt and become dominant and unmanageable. In practical terms this means as diverse and long a rotation as possible consistent with the farm system and which prevents the weeds returning seeds to the soil seed bank.

Crop Rotation

Crop planning is a cornerstone of organic farming practice and it has important implications for weed management. It can be designed to positively influence weed control and to make a useful contribution to the whole farm management strategy. Typically rotation cycles extend over several years with often only an annual change of crop, but the inclusion of cover crops, intercrops and green manures increases the crop diversity in a rotation. In horticultural systems there may be sequential cropping where short-term crops follow each other in succession.

Weed population density may be markedly reduced using crop rotation but there has been little experimentation. Success depends on the use of crop sequences that create a diverse pattern of competition, allelopathic interference, soil disturbance, and production needs (such as the time of sowing and harvesting). There should be regular changes between spring and autumn-sown crops, and between annual and perennial crops, between dense leafy crops and those with an open habit, and between crops that require a long growing season and others that mature quickly. Rotation may also allow the use of a range of cultivations and direct non-chemical weeding methods that may be applicable to the different crops. The aim is to provide an unstable and inhospitable environment that prevents the proliferation of a particular weed species.

Choosing Crops and their Sequence

The length of the rotation, the choice and sequence of crops will depend upon individual farming circumstances that will include factors like soil type, rainfall, topography and enterprises. However, the aim is to produce an unstable environment in which no single weed species is allowed to adapt, become dominant, and therefore difficult to manage. No one rotation can be recommended, but ideally in terms of weed control rotations should include:

• Alternation of autumn and spring germinating crops,

- Alternation of annual and perennial crops (including grass),

- Alternation of closed, dense crops such as oats which shade out weeds, and open crops such as maize which encourage weeds,

- A variety of cultivations and cutting or topping operations that directly affect the weeds.

Various suggestions and observations include:

- Putting sensitive annual crops after perennial leys. Research has shown that in the third cropping year after a grass/clover ley there is twice as much weed emergence as compared to the first.

- Include a row crop that allows the use of one or more cultivations to kill emerged weeds and encourages the germination of others, so reducing the soil seedbank and hence potential weed numbers in future crops. Cultivations may also reduce the problem of perennial weeds by disrupting growth and smothering regeneration in the growing crop. Typical cleaning crops include turnip, sugar beet, and potato.

- Uncultivated leys provide a completely different habitat for weeds and may be used to reduce or eliminate particular weed species. Few studies have been made of the effectiveness of leys for controlling weeds but trials suggest that there is little advantage for weed management in leaving leys down longer than 3 years. The species composition, and the mowing and grazing regimes are important in the weed dynamics. Management of the weeds at the time of ley establishment is critical as is the method of ending the ley to avoid a flush of weeds due to the release of seed dormancy by cultivation. A greater proportion of ley in the rotation usually results in lower seed numbers in the seedbank in comparison with arable crops. It was a traditional way to deal with land infested with wild oat but does not eliminate the weed completely.

- Where a long grass break does not form part of the rotation weed problems are likely to be more severe. The problem will be greater where less vigorous and therefore less competitive crops are grown. Among the cereals, oats and winter rye are the most competitive followed by triticale.

- Canopy development and shading are important for weed suppression and choice of cultivar can influence this.

- Higher seeding rate and narrower row spacing increase the level of weed suppression competitive cereals like rye may be grown as short duration.

Fertility Building

The fertility-building period, or ley, will influence the weed population. If it is well managed it can act as a weed suppressing phase. It is important to choose the right species and ensure they establish quickly, especially for grassland systems where the ley may last for several years. Establishment of leys can be easier in the autumn period than in the spring because sowing in spring coincides with the main spring flush of weeds. The seedbed needs to be well prepared, and good contact made between seed mix and, ideally, moist soil to achieve good establishment. The choice of fertility building crop is also important. Rotations with grass leys have been shown to be beneficial in reducing weed seed numbers compared with rotations that do not include a grass phase. Grassland systems, which have temporary leys rather than permanent pasture, will provide the opportunity to control perennial weeds during the cultivations between ploughing and reseeding.

Grassland or clover/grass leys are an important part of the organic farming system in the UK. On livestock farms grassland forms the basis of the production process, in arable systems the ley is used primarily for maintaining or restoring soil fertility. The grass may be managed as a short, medium or long-term crop and this may determine the composition of the desirable sward species and the nature of the associated weeds. The seed mixture for a ley may include a relatively simple mixture of grasses and legumes or may be more complex and contain a range.

Choosing Varieties

Varieties that consistently suppress weeds are generally more desirable in organic systems (although this might be outweighed by marketing necessities) as opposed to varieties that tolerate weeds (and which potentially allow weeds to develop and return seeds to the soil seed bank).

Organic varieties: Should show quick germination and establishment, rapid early and vigorous growth, and the ability to rapidly cover the soil and shade it (prostrate or tall varieties) to out-compete weeds at an early a stage in the crop cycle as possible. Varieties are well known to differ in architecture and competitive ability and whilst those that out-compete weeds are preferred it should also be borne in mind that those with erect foliage or that can tolerate some degree of mechanical weeding are also likely to be useful. Some crop types or varieties might produce allelochemicals that prevent weeds from developing or germinating although information on this is generally lacking.

In grassland systems the choice of variety may be dominated by forage value, but if there is opportunity the most vigorous species should be selected, as these will determine the productivity of the whole ley period. The trend in conventional cereal production has been to grow the taller stemmed varieties for their weed suppressing ability. Some farmers have stayed with the shorter stemmed varieties and employed a weed topper/cutter which will remove and, ideally, collect weed seed-heads so long as there is a difference in height between crop and weed. New research in wheat is investigating leaf angle development, height and speed of development on weed suppression to aid farmer variety choice.

Seed size: Varieties with a larger seed size have been shown to exhibit greater initial vigour of emergence and growth, which may subsequently provide extra competitive ability. If there is a choice available then the most vigorous species should be selected, as these will be more likely to out-compete weeds and suppress their development.

Clean seed: It is important that the crop seed is free from contamination by weed seeds. Organic farmers are obliged to use organically produced seed and this should be clean. It is important if saving seed that it is taken from weed free crops, and ideally, professionally cleaned. Tolerance levels of contamination should be low although they are not generally well defined as of yet. Tolerance levels of dock in Switzerland are one seed per 100 g.

Seed Rate and Crop Spacing

Spatial distribution of the canopy foliage and rooting system will be important for weed suppression. In drilled or transplanted crops the proximity of the plants to one another will determine the competitiveness of the plant stand as a whole. The principle is that the greater the amount of space taken up by the crop in the rows, the less space there is available for the weeds to invade. However, it should be borne in mind that closely spaced crop plants compete with each other and that it is also expedient to allow sufficient space between plants to allow for efficient mechanical weeding should weeds develop and threaten the crop.

Seed rates tend to be higher for organic than conventional crops. There is also an allowance for potentially lower germination rates and loss of the crop by mechanical weeders.

There has been much work in cereals on row spacing, pattern, direction of sowing and seed rates (typically 10% higher in organic cereals). Results are varied and interactions between the factors are often significant. For example in narrow widths an E-W sowing was favourable whereas in wide row widths a N-S sowing showed better response. Findings from the EU funded WECOF trials are awaited to give more definite recommendations.

In vegetable crop market size specifications are often the main driver in determining the crop row spacing.

Establishing the Crop

The ability of the crop to get off to a good start ahead of the weed flora is critical. Good soil management practises are important to provide the best possible seedbed in which to plant a crop. The impact of a poor compacted soil can soon be seen on crop establishment and subsequent weed invasion.

In some systems sowing can be aided by the use of primed seed, or by transplanting an already established plant into a freshly prepared weed free seedbed.

Transplanting is a popular technique in organic horticultural systems. Bare rooted transplants can be raised on holdings or modular plants raised or bought in then planted out in the field. Advantages include the benefit of accurate spacing i.e. not having to rely on germination that can sometimes lead to uneven establishment with subsequent yield and quality penalties. It also accentuates the difference in size between crop and weed, which can be vital for mechanical weeding at later stages.

Intercropping and Undersowing

Intercropping (or mixed cropping) and undersowing involves growing two or more different crop species in the same area. The advantage for weed control is that the crops cover more ground, so there is less space available for weed emergence. Intercropping can involve purely cash crops or a mixture of cash crops and fertility building crops.

Intercropping: Is widely practised in certain countries and an enormous variety of intercropping systems have been developed. Both component crops may be taken to yield or one may be there as a living mulch to improve weed control. In successful intercrops weed suppression is usually superior to that of either of the component crops when grown alone. Crop density, crop diversity, crop spatial arrangement, choice of crop species and cultivar will all affect weed growth in intercropping systems. If water or nutrients are limiting then growth of one or both intercrops will suffer. Improved weed control alone is unlikely to justify their use and there must be other obvious benefits if the change in cropping practice is to prove economic.

In terms of mixed cash cropping there have been investigations into organic winter wheat and beans that reduced weed growth and gave better yields than sole cropping. Leeks and celery intercropping has also been shown to increase weed suppressing ability, and reduce reproductive capacity of late emerging groundsel. There are probably a wide range of combinations that could be designed to suit the farmer's rotations and marketing needs although some practical experimentation is required.

Undersowing: Aims to cover the ground with a quick-growing dense layer of vegetation underneath the crop. The undersown species is prostrate, usually leguminous, and adds to or maintains fertility. It also suppresses weeds. Combining cash cropping with fertility building in this manner potentially produces an economic return, and it may mean there is either no need for an isolated fertility period, or that the length of that phase can be reduced. Undersowing cash crops with fertility building crops has other advantages apart from weed suppression, but so far the technique has only been widely used in cereal growing where it helps to re-establishing leys and avoid bare ground after harvest. Short-strawed cereal varieties can be difficult to manage, and straw difficult to save, due to the undersown crop growing up into the cereal. Further research into variety choice and sowing rates is needed to ensure that competition between the two crops is kept to a minimum.

Using Cutting Regimes for Weed Control

Cutting and topping weeds will have an impact on the type and weed flora in a field and can be invaluable in preventing return of weed seed to the soil seed bank. Cutting and topping are important for weed management in pasture, grass and leys. Topping can also be used as a remedial measure in vegetable or other crops to prevent weeds from seeding.

Pasture systems: Good management involves maintaining the condition of the sward by cultural means. In particular reducing weed intrusions by chain harrowing in spring and topping regularly during the growing season. Mowing during the seeding year must be carefully judged and close cutting avoided. Spring sown stands should be cut no later than mid-August to allow recovery before winter. Summer sowings should be left unmown until November. Undersown lucerne should be left to grow into the winter. Where companion grasses are growing strongly, light winter grazing may be desirable. In grazed pasture weeds that are not eaten by livestock, will need to be topped to prevent seed shed. The established crop may be cut up to four times per year starting in mid-May. The crop is quickly weakened by defoliation, either by grazing or cutting, at too young a stage especially in spring or autumn. Before entering the dormant stage the crop must be allowed to make sufficient growth to replenish the food reserves in the root.

Grassland systems: Cutting for hay or silage will have an impact on the weed flora. Silage tends to be cut early in the season when the sward is young and fresh, whilst hay is cut at a later stage. There can be both advantages and disadvantages associated with the timing of cutting depending on the weed flora and the ultimate requirements of the system. Cutting late may allow weeds in the pasture to grow to maturity and set seed. The ripe seeds may contaminate the hay and remain viable when passed through livestock. Dock seeds should not survive low pH silage, however they will survive in a later cut of hay. This mature seed may also shed on the ley surface and find opportunities to germinate in situ or be transported by livestock to other locations. In contrast, cutting early for silage in fields, with for example an infestation of creeping thistle, may encourage the spread and growth of this weed. Hence, there has to be a balance between the requirements of the farming system and weed control implications.

Horticultural and stockless arable systems: Ley management will include topping at intervals during the summer to a height of around 10-15 cm. Ideally in fertility building leys the sward should not be allowed to get higher than 40 cm (or knee height). If the vegetation gets higher than this, then topping will create a mat of vegetation that will act like a mulch. This can create dead spots in the ley where clover may be excluded by the more vigorous grasses, or which weeds may colonise. Topping the ley regularly will also ensure that tall weeds that may have germinated will not be able set seed.

Using Manures

The use of raw manures and slurry has often been associated by farmers with increased weed problems. The problems can arise in various ways; either as a result of weed seeds in the manure, as a result of the way in which it is applied or due to the stimulatory effect of the nutritents on weeds already present in the soil.

Weed seeds in manure: Some manure contains weed seed, either seed that has passed undigested through animals or from bedding materials like small-grain straw and old hay. High-temperature aerobic composting (recommended under organic standards) can greatly reduce the number of viable weed seeds as long as the temperature is maintained at higher than 60 °C for more than three days. Operationally compost will need to be regularly turned to achieve even heating through the whole heap and to get material from the outside (where seeds are likely to survive) to the inside (where the highest temperatures are likely to be generated. In a similar way, aeration of slurry can reduce the number and viability of weed seeds.

Applying manure: When applying manure or slurry try not to create conditions which stimulate weed seeds to germinate (excessive soil disturbance, creating bare patches etc). Applying slurry to stubble after silage cuts can provide optimum conditions for weed seed germination. A nutrient-rich bed of cattle slurry will produce a high potassium environment which will favour weeds such as docks rather than grasses. Dock seeds should not survive low ph silage but will survive in a later cut of hay.

Some research has shown that placing manures and slurry more accurately on crops can benefit the crop rather than the weeds. Crop plants are generally sown fairly deeply and they germinate from a lower level in the soil profile than weeds, which tend to dwell on the surface and germinate from 0-3 cm. Crop plants also root more deeply. This tendency can be exploited for weed management. In arable/horticultural systems manure placed 10 cm below the soil surface encourages the crop seeds to grow down into the nutrient-rich layer before the surface-dwelling weeds can reach it. This technique can also be used with broadcast spread slurry. If it is ploughed in rather than left on the surface it will be available to the crop before the weeds can reach it.

In many cases, the growth of weeds that follows manuring is a result of the stimulating effect manure has on weed seeds already present in the soil. This can be due to the flush of nutrients (e.g. supply of nitrates), enhanced biological activity in the soil, or other changes in the fertility status of the soil. Some work has indicated that excesses of potash and nitrogen in particular can encourage weeds but in any case it is prudent to monitor the nutrient content of your soil and manure, and spread manure evenly to reduce the incidence of weed problems.

Don't give the docks an advantage: applying slurry to stubble after silage cuts can provide optimum conditions for weed seed germination. A nutrient-rich bed of cattle slurry will produce a high potassium environment which will favour weeds such as docks rather than grasses. Dock seeds should not survive low ph silage but will survive in a later cut of hay.

Weed Management and Livestock

In mixed systems, where grass/clover leys are used for fertility building, livestock can make good use of the nutrients and they also produce manure, a resource which can be used around the farm to fertilise cash crops. Apart from leys, pastures will also need weed management and increasingly conservation of old or rough pastures requires specialist grazing. Animals can also be used to consume cut weeds or other plant material like chaff or screenings that are likely to contain some weed seeds.

Animals have different grazing habits and it is even recognised that different breeds or individuals are likely to have different tastes and habits. The species, breed, age and individuality of animals will all affect what they will eat and therefore what effect they will have on both weeds and pasture. Variability within the feeding site (e.g. vegetation, topography) can also be important as can other factors such as the weather. In general terms:

- Goats are browsers and have a reputation for enjoying tough and woody plants.

- Sheep are recognised as being useful for weed control as they graze close to the ground and will eat a wide range of plants. They can be used early and some breeds are hardy.

- Cattle can be used for early grazing but there are a large number of different breeds and types with different grazing requirements including beef, dairy and traditional breeds. Grazing strategies appear to be related to plant energy content and digestibility and this will affect how plants are eaten (leaves or stems or other parts of plants), which plants are eaten (species) and size plants eaten (young or more mature plants). Cattle tend to avoid longer coarser grass and hairy, spiny or poisonous plants. The selection of certain plant species and plant components as well as the location of these plants is based on the previous experience of the animals or learned from their mothers when they are calves.

- Pigs are good at rooting and have been recommended at various times for digging out perennial weeds like dock and couch when fenced within fields (and tightly stocked).

- Geese consume grassy weeds and have been used to weed in between rows of well established crops.

- Horses and ponies are grazed on ever increasing areas of land and can be used as part of a grazing rotation. They prefer frequent small amounts of fibrous grass or other high roughage material. They have been known to dramatically increase the number of docks in a rotation.

It is important to get the right grazing balance over the year to get the maximum benefit for the animals and also to prevent damage to the sward or soil. For example, stocking more lightly in the winter months and in wet periods prevents poaching. So think about the right season to graze, how long to graze, how many animals to graze and how long the grazing area will need to recover. Things to consider include:

- Timing grazing to benefit the pasture and promote competition with weeds.

- Timing grazing to damage the weeds, e.g. to remove flowers or seed heads before seed production.

- Allowing time for the pasture or forage to recover between grazings.

- Making sure that livestock that have been grazing on weedy land feed on weed seed free forage for 4-5 days before introducing them to weed free areas or pastures (some seeds will remain viable after passing through animals which may take a few days).

Suggestions for rotating livestock, depending on situation, include:

- Alternate the grazing of sheep and cattle from year to year or to use mixed grazing for better weed control. Mixed grazing in the same field may be detrimental to the cattle.

- Exploit animals.

Fallowing

It is usually not desirable to have to plan a fallow period into a rotation, but it may be necessary if weeds cannot be controlled during cropping or fertility building. It may not be necessary to stop cropping for a whole year, but instead to employ a bastard fallow i.e. no crop for part of the year. Tillage without a crop for a season is sometimes referred to as a black fallow. Fallowing the land for part of the growing season, as a bastard or summer fallow, can be as effective as a full fallow, is more suitable for lighter land and can be fitted into most rotations.

Fallowing is often best during the summer when cultivations can take place and the drier periods allow for root desiccation. This technique is more useful in plough-dominated systems rather than grassland management. One aim is to cultivate the soil progressively deeper over time, exposing underground plant parts to desiccation at the soil surface but in this case dry weather conditions are essential. Ploughing begins in June/July allowing time for an early crop to precede it. A bastard fallow is often used after a ley to reduce perennial weeds before sowing a winter cereal. There is also an opportunity for birds to feed on wireworms exposed during soil disturbance.

Fallowing has been shown to reduce perennial weeds within a rotation. The aim is to kill the vegetative organs of the weeds by mechanical damage and desiccation. For a full or bare fallow, heavy land is ploughed in April to give the weeds time to start into growth. It is cultivated or cross-ploughed 10-14 days later to produce a cloddy tilth. The soil is cultivated or ploughed at frequent intervals to move the clods around and dry them out. By August the clods should have broken down and the soil is left to allow the weed seeds to germinate. In September/October the weeds are ploughed in and the land prepared for autumn cropping. If a cereal is to follow the fallow, wheat bulb fly may be a problem because it lays eggs on bare ground in July. This can be overcome by sowing a green manure such as mustard to cover the land during this period.

Although there is the benefit of reduced weed control costs in subsequent crops after an effective fallow, the economics of taking land out of production for a full year together with undesirable effects on the soil and the environment, make the use of a bare fallow unlikely for weed control in the organic system. There is no financial return during the fallow period while labour costs accumulate during the fallowing operations. As an alternative to fallowing, cleaning crops such as potatoes and turnips allow repeated hoeing for weed control (but are not suited to heavy land).

A similar effect to that of fallowing can be achieved with rapidly developing crops like radish (Raphanus sativus) that are harvested before the onset of weed competition. The short interval between crop establishment and harvesting in this crop encourages weed seed germination but does not allow the weeds time to set seed or reproduce vegetatively.

Farm Hygiene

Weeds are, by their nature tenacious and almost impossible to eradicate once established. The best form of management is preventing their establishment in the first place. Weeds are easily spread between fields and between farms and it is worth taking some trouble to try and prevent this with some basic hygiene measures. Ask yourself the following questions:

Have you got a system for detecting weeds early?

- Managing a particular weed will be easier if it is detected early and prevented from spreading.

- Ensure that all people who work on the farm or visit it are alert to the possibility of spreading weeds and weed seed and ask them to tell you if they notice any particular areas of weeds.

- Keep records of problems weeds and their spread or otherwise. A digital camera can be a useful tool to record presence of weeds and monitor changes over time.

Is your farm machinery spreading weeds?

- Weed seeds are easily carried in soil, crop residues and on machinery so these should be regularly cleaned down.

- If there is a serious weed infestation in a particular field, or machinery is moving through fields where weeds are flowering, then washing down machinery should be a serious consideration.

- Hygiene is particularly important at harvest time. In crops like cereals weed seeds may be scattered in the field or caught on the machinery and dislodged later some distance from the original source. It may be necessary to add screens to combines to catch weed seeds at harvest. Older models may already have these features.

Tillage

Cultivating after an early rain.

Tillage is the agricultural preparation of soil by mechanical agitation of various types, such as digging, stirring, and overturning. Examples of human-powered tilling methods using hand tools include shovelling, picking, mattock work, hoeing, and raking. Examples of draft-animal-powered or mechanized work include ploughing (overturning with moldboards or chiseling with chisel shanks), rototilling, rolling with cultipackers or other rollers, harrowing, and cultivating with cultivatorshanks (teeth). Small-scale gardening and farming, for household food production or small businessproduction, tends to use the smaller-scale methods, whereas medium- to large-scale farming tends to use the larger-scale methods.

Tillage that is deeper and more thorough is classified as primary, and tillage that is shallower and sometimes more selective of location is secondary. Primary tillage such as ploughing tends to produce a rough surface finish, whereas secondary tillage tends to produce a smoother surface finish, such as that required to make a good seedbed for many crops. Harrowing and rototilling often combine primary and secondary tillage into one operation.

"Tillage" can also mean the land that is tilled. The word "cultivation" has several senses that overlap substantially with those of "tillage". In a general context, both can refer to agriculture. Within agriculture, both can refer to any kind of soil agitation. Additionally, "cultivation" or "cultivating" may refer to an even narrower sense of shallow, selective secondary tillage of row crop fields that kills weeds while sparing the crop plants.

Primary and Secondary Tillage

Primary Tillage is usually conducted after the last harvest, when the soil is wet enough to allow plowing but also allows good traction. Some soil types can be plowed dry. The objective of primary tillage is to attain a reasonable depth of soft soil, incorporate crop residues, kill weeds, and to aerate the soil. Secondary tillage is any subsequent tillage, in order to incorporate fertilizers, reduce the soil to a finer tilth, level the surface, or control weeds.

Reduced Tillage

Plough tilling the field.

Reduced tillage leaves between 15 and 30% crop residue cover on the soil or 500 to 1000 pounds per acre (560 to 1100 kg/ha) of small grain residue during the critical erosion period. This may involve the use of a chisel plow, field cultivators, or other implements.

Intensive Tillage

Intensive tillage leaves less than 15% crop residue cover or less than 500 pounds per acre (560 kg/ha) of small grain residue. This type of tillage is often referred to as conventional tillage but as conservational tillage is now more widely used than intensive tillage (in the United States), it is often not appropriate to refer to this type of tillage as conventional. Intensive tillage often involves multiple operations with implements such as a mold board, disk, and/or chisel plow. Then a finisher with a harrow, rolling basket, and cutter can be used to prepare the seed bed. There are many variations.

Conservation Tillage

Conservation tillage leaves at least 30% of crop residue on the soil surface, or at least 1,000 lb/ac (1,100 kg/ha) of small grain residue on the surface during the critical soil erosion period. This slows water movement, which reduces the amount of soil erosion. Additionally, conservation tillage has been found to benefit predatory arthropods that can enhance pest control. Conservation tillage also benefits farmers by reducing fuel consumption and soil compaction. By reducing the number of times the farmer travels over the field, farmers realize significant savings in fuel and labor. In most years since 1997, conservation tillage was used in US cropland more than intensive or reduced tillage.

However, conservation tillage delays warming of the soil due to the reduction of dark earth exposure to the warmth of the spring sun, thus delaying the planting of the next year's spring crop of corn.

- No-till - Never use a plow, disk, etc. ever again. Aims for 100% ground cover.

- Strip-Till - Narrow strips are tilled where seeds will be planted, leaving the soil in between the rows untilled.

- Mulch-till

- Rotational Tillage - Tilling the soil every two years or less often (every other year, or every third year, etc).

- Ridge-Till.

Zone Tillage

Zone tillage is form of modified deep tillage in which only narrow strips are tilled, leaving soil in between the rows untilled. This type of tillage agitates the soil to help reduce soil compaction problems and to improve internal soil drainage. It is designed to only disrupt the soil in a narrow strip directly below the crop row. In comparison to no-till, which relies on the previous year's plant residue to protect the soil and aides in postponement of the warming of the soil and crop growth in Northern climates, zone tillage creates approximately a 5-inch-wide strip that simultaneously breaks up plow pans, assists in warming the soil and helps to prepare a seedbed. When combined with cover crops, zone tillage helps replace lost organic matter, slows the deterioration of the soil, improves soil drainage, increases soil water and nutrient holding capacity, and allows necessary soil organisms to survive.

It has been successfully used on farms in the mid-west and west for over 40 years and is currently used on more than 36% of the U.S. farmland. Some specific states where zone tillage is currently in practice are Pennsylvania, Connecticut, Minnesota, Indiana, Wisconsin, and Illinois.

Unfortunately, there aren't consistent yield results in the Northern Cornbelt states; however, there is still interest in deep tillage within the agriculture industry. In areas that are not well-drained, deep tillage may be used as an alternative to installing more expensive tile drainage.

Effects of Tillage

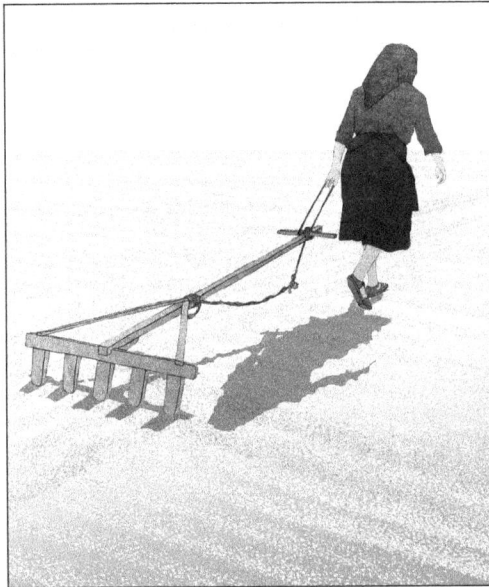

Rice tillage.

Positive

Plowing:

- Loosens and aerates the top layer of soil or horizon A, which facilitates planting the crop.

- Helps mix harvest residue, organic matter (humus), and nutrients evenly into the soil.

- Mechanically destroys weeds.

- Dries the soil before seeding (in wetter climates tillage aids in keeping the soil drier).

- When done in autumn, helps exposed soil crumble over winter through frosting and defrosting, which helps prepare a smooth surface for spring planting.

Negative

- Dries the soil before seeding.

- Soil loses a lot of nutrients, like nitrogen and fertilizer, and its ability to store water.

- Decreases the water infiltration rate of soil. (Results in more runoff and erosion since the soil absorbs water slower than before)

- Tilling the soil results in dislodging the cohesiveness of the soil particles thereby inducing erosion.

- Chemical runoff.

- Reduces organic matter in the soil.

- Reduces microbes, earthworms, ants, etc.

- Destroys soil aggregates.

- Compaction of the soil, also known as a tillage pan.

- Eutrophication (nutrient runoff into a body of water).

- Can attract slugs, cut worms, army worms, and harmful insects to the left over residues.

- Crop diseases can be harbored in surface residues.

General Comments

- The type of implement makes the most difference, although other factors can have an effect.

- Tilling in absolute darkness (night tillage) might reduce the number of weeds that sprout following the tilling operation by half. Light is necessary to break the dormancy of some weed species' seed, so if fewer seeds are exposed to light during the tilling process, fewer will sprout. This may help reduce the amount of herbicides needed for weed control.

- Greater speeds, when using certain tillage implements (disks and chisel plows), lead to more intensive tillage (i.e., less residue is on the soil surface).

- Increasing the angle of disks causes residues to be buried more deeply. Increasing their concavity makes them more aggressive.

- Chisel plows can have spikes or sweeps. Spikes are more aggressive.

- Percentage residue is used to compare tillage systems because the amount of crop residue affects the soil loss due to erosion.

Alternatives to Tilling

Modern agricultural science has greatly reduced the use of tillage. Crops can be grown for several years without any tillage through the use of herbicides to control weeds, crop varieties that tolerate packed soil, and equipment that can plant seeds or fumigate the soil without really digging it up. This practice, called no-till farming, reduces costs and environmental change by reducing soil erosion and diesel fuel usage.

Site Preparation of Forest Land

Site preparation is any of various treatments applied to a site in order to ready it for seeding or planting. The purpose is to facilitate the regeneration of that site by the chosen method. Site preparation may be designed to achieve, singly or in any combination: improved access, by reducing or rearranging slash, and amelioration of adverse forest floor, soil, vegetation, or other biotic factors. Site preparation is undertaken to ameliorate one or more constraints that would otherwise be likely to thwart the objectives of management. A valuable bibliography on the effects of soil temperature and site preparation on subalpine and boreal tree species has been prepared by McKinnon et al.

Site preparation is the work that is done before a forest area is regenerated. Some types of site preparation are burning.

Burning

Broadcast burning is commonly used to prepare clearcut sites for planting, e.g., in central British Columbia, and in the temperate region of North America generally.

Prescribed burning is carried out primarily for slash hazard reduction and to improve site conditions for regeneration; all or some of the following benefits may accrue:

- Reduction of logging slash, plant competition, and humus prior to direct seeding, planting, scarifying or in anticipation of natural seeding in partially cut stands or in connection with seed-tree systems.

- Reduction or elimination of unwanted forest cover prior to planting or seeding, or prior to preliminary scarification thereto.

- Reduction of humus on cold, moist sites to favour regeneration.

- Reduction or elimination of slash, grass, or brush fuels from strategic areas around forested land to reduce the chances of damage by wildfire.

Prescribed burning for preparing sites for direct seeding was tried on a few occasions in Ontario, but none of the burns was hot enough to produce a seedbed that was adequate without supplementary mechanical site preparation.

Changes in soil chemical properties associated with burning include significantly increased pH, which Macadam in the Sub-boreal Spruce Zone of central British Columbia found persisting more than a year after the burn. Average fuel consumption was 20 to 24 t/ha and the forest floor depth was reduced by 28% to 36%. The increases correlated well with the amounts of slash (both total and ≥7 cm diameter) consumed. The change in pH depends on the severity of the burn and the amount consumed; the increase can be as much as 2 units, a 100-fold change. Deficiencies of copper and iron in the foliage of white spruce on burned clearcuts in central British Columbia might be attributable to elevated pH levels.

Even a broadcast slash fire in a clearcut does not give a uniform burn over the whole area. Tarrant, for instance, found only 4% of a 140-ha slash burn had burned severely, 47% had burned lightly, and 49% was unburned. Burning after windrowing obviously accentuates the subsequent heterogeneity.

Marked increases in exchangeable calcium also correlated with the amount of slash at least 7 cm in diameter consumed. Phosphorus availability also increased, both in the forest floor and in the 0 cm to 15 cm mineral soil layer, and the increase was still evident, albeit somewhat diminished, 21 months after burning. However, in another study in the same Sub-boreal Spruce Zone found that although it increased immediately after the burn, phosphorus availability had dropped to below pre-burn levels within 9 months.

Nitrogen will be lost from the site by burning, though concentrations in remaining forest floor were found by Macadam to have increased in 2 of 6 plots, the others showing decreases. Nutrient losses may be outweighed, at least in the short term, by improved soil microclimate through the reduced thickness of forest floor where low soil temperatures are a limiting factor.

The *Picea/Abies* forests of the Alberta foothills are often characterized by deep accumulations of organic matter on the soil surface and cold soil temperatures, both of which make reforestation difficult and result in a general deterioration in site productivity; Endean and Johnstone describe experiments to test prescribed burning as a means of seedbed preparation and site amelioration on representative clear-felled *Picea/Abies* areas. Results showed that, in general, prescribed burning did not reduce organic layers satisfactorily, nor did it increase soil temperature, on the sites tested. Increases in seedling establishment, survival, and growth on the burned sites were probably the result of slight reductions in the depth of the organic layer, minor increases in soil temperature, and marked improvements in the efficiency of the planting crews. Results also suggested that the process of site deterioration has not been reversed by the burning treatments applied.

Ameliorative Intervention

Slash weight (the oven-dry weight of the entire crown and that portion of the stem <4 inches in diameter) and size distribution are major factors influencing the forest fire hazard on harvested sites. Forest managers interested in the application of prescribed burning for hazard reduction and silviculture, were shown a method for quantifying the slash load by Kiil. In west-central Alberta, he felled, measured, and weighed 60 white spruce, graphed (a) slash weight per merchantable unit volume against diameter at breast height (dbh), and (b) weight of fine slash (<1.27 cm) also against dbh, and produced a table of slash weight and size distribution on one acre of a hypothetical stand of white spruce. When the diameter distribution of a stand is unknown, an estimate of slash weight and size distribution can be obtained from average stand diameter, number of trees per unit area, and merchantable cubic foot volume. The sample trees in Kiil's study had full symmetrical crowns. Densely growing trees with short and often irregular crowns would probably be overestimated; open-grown trees with long crowns would probably be underestimated.

The need to provide shade for young outplants of Engelmann spruce in the high Rocky Mountains is emphasized by the U.S. Forest Service. Acceptable planting spots are defined as microsites on the north and east sides of down logs, stumps, or slash, and lying in the shadow cast by such material. Where the objectives of management specify more uniform spacing, or higher densities, than obtainable from an existing distribution of shade-providing material, redistribution or importing of such material has been undertaken.

Access

Site preparation on some sites might be done simply to facilitate access by planters, or to improve access and increase the number or distribution of microsites suitable for planting or seeding.

Wang et al. determined field performance of white and black spruces 8 and 9 years after outplanting on boreal mixedwood sites following site preparation (Donaren disc trenching versus no trenching) in 2 plantation types (open versus sheltered) in southeastern Manitoba. Donaren trenching slightly reduced the mortality of black spruce but significantly increased the mortality of white spruce. Significant difference in height was found between open and sheltered plantations for black spruce but not for white spruce, and root collar diameter in sheltered plantations was significantly larger than in open plantations for black spruce but not for white spruce. Black spruce open plantation had significantly smaller volume ($97 \, cm^3$) compared with black spruce sheltered ($210 \, cm^3$), as well as white spruce open ($175 \, cm^3$) and sheltered ($229 \, cm^3$) plantations. White spruce open plantations also had smaller volume than white spruce sheltered plantations. For transplant stock, strip plantations had a significantly higher volume ($329 \, cm^3$) than open plantations ($204 \, cm^3$). Wang et al. recommended that sheltered plantation site preparation should be used.

Mechanical

Up to 1970, no "sophisticated" site preparation equipment had become operational in Ontario, but the need for more efficacious and versatile equipment was increasingly recognized. By this time, improvements were being made to equipment originally developed by field staff, and field testing of equipment from other sources was increasing.

According to J. Hall, in Ontario at least, the most widely used site preparation technique was post-harvest mechanical scarification by equipment front-mounted on a bulldozer (blade, rake, V-plow, or teeth), or dragged behind a tractor (Imsett or S.F.I. scarifier, or rolling chopper). Drag type units designed and constructed by Ontario's Department of Lands and Forests used anchor chain or tractor pads separately or in combination, or were finned steel drums or barrels of various sizes and used in sets alone or combined with tractor pad or anchor chain units.

J. Hall's report on the state of site preparation in Ontario noted that blades and rakes were found to be well suited to post-cut scarification in tolerant hardwood stands for natural regeneration of yellow birch. Plows were most effective for treating dense brush prior to planting, often in conjunction with a planting machine. Scarifying teeth, e.g., Young's teeth, were sometimes used to prepare sites for planting, but their most effective use was found to be preparing sites for seeding, particularly in backlog areas carrying light brush and dense herbaceous growth. Rolling choppers found application in treating heavy brush but could be used only on stone-free soils. Finned drums were commonly used on jack pine–spruce cutovers on fresh brushy sites with a deep duff layer and heavy slash, and they needed to be teamed with a tractor pad unit to secure good distribution of the slash. The S.F.I. scarifier, after strengthening, had been "quite successful" for 2 years, promising trials were under way with the cone scarifier and barrel ring scarifier, and development had begun on a new flail scarifier for use on sites with shallow, rocky soils. Recognition of the need to become more effective and efficient in site preparation led the Ontario Department of Lands and Forests to adopt the policy of seeking and obtaining for field testing new equipment from Scandinavia and elsewhere that seemed to hold promise for Ontario conditions, primarily in the north. Thus, testing was begun of the Brackekultivator from Sweden and the Vako-Visko rotary furrower from Finland.

Mounding

Site preparation treatments that create raised planting spots have commonly improved outplant performance on sites subject to low soil temperature and excess soil moisture. Mounding can certainly have a big influence on soil temperature. Draper et al., for instance, documented this as well as the effect it had on root growth of outplants.

The mounds warmed up quickest, and at soil depths of 0.5 cm and 10 cm averaged 10 and 7 °C higher, respectively, than in the control. On sunny days, daytime surface temperature maxima on the mound and organic mat reached 25 °C to 60 °C, depending on soil wetness and shading. Mounds reached mean soil temperatures of 10 °C at 10 cm depth 5 days after planting, but the control did not reach that temperature until 58 days after planting. During the first growing season, mounds had 3 times as many days with a mean soil temperature greater than 10 °C than did the control microsites.

Draper et al.'s mounds received 5 times the amount of photosynthetically active radiation (PAR) summed over all sampled microsites throughout the first growing season; the control treatment consistently received about 14% of daily background PAR, while mounds received over 70%. By November, fall frosts had reduced shading, eliminating the differential. Quite apart from its effect on temperature, incident radiation is also important photosynthetically. The average control microsite was exposed to levels of light above the compensation point for only 3 hours, i.e., one-quarter of the daily light period, whereas mounds received light above the compensation point for 11 hours, i.e., 86% of the same daily period. Assuming that incident light in the 100-600 $\mu Em^{-2}s^{-1}$ intensity range is the most important for photosynthesis, the mounds received over 4 times the total daily light energy that reached the control microsites.

Orientation of Linear Site Preparation

With linear site preparation, orientation is sometimes dictated by topography or other considerations, but the orientation can often be chosen. It can make a difference. A disk-trenching experiment in the Sub-boreal Spruce Zone in interior British Columbia investigated the effect on growth of young outplants (lodgepole pine) in 13 microsite planting positions: berm, hinge, and trench in each of north, south, east, and west aspects, as well as in untreated locations between the furrows. Tenth-year stem volumes of trees on south, east, and west-facing microsites were significantly greater than those of trees on north-facing and untreated microsites. However, planting spot selection was seen to be more important overall than trench orientation.

In a Minnesota study, the N–S strips accumulated more snow but snow melted faster than on E–W strips in the first year after felling. Snow-melt was faster on strips near the centre of the strip-felled area than on border strips adjoining the intact stand. The strips, 50 feet (15.24 m) wide, alternating with uncut strips 16 feet (4.88 m) wide, were felled in a *Pinus resinosa* stand, aged 90 to 100 years.

Crop Weed Management Strategies

Amenity and Conservation Areas

Areas of amenity land are found on roadside verges, along railways, grassed areas on riverbanks, golf course roughs, in cemeteries, country and urban parks and on land associated with housing and landscaped areas around other building complexes. The planting may be formal or not depending on the location but visual appearance is often of some importance. Grasses make up the greatest proportion of the vegetation in amenity and conservation sites whether as a sports or playing surface or simply as ground cover. Whatever the vegetation some form of maintenance will be required and it is likely that the cost of doing so must be kept to a minimum.

While amenity plantings are not considered a crop where yield needs to be optimised, the desired plants do need to be protected from weed competition, at least during establishment. For instance, newly planted trees and shrubs can suffer severe stunting if left unweeded. However, there is increasing interest in using amenity areas to serve as refuges for wild plants and animals. Certain areas such as field boundaries, hedges and ditches are of particular importance for wild life conservation.

Apart from maintaining biodiversity, the objectives of management may include height control, maintenance of species composition, weed control to limit certain species, reduction of fire risk or some requirement specific to the location.

Field Vegetables

Alliums

Allium crops germinate and grow slowly, they generally have rather few leaves and for most of the growing season the soil is not fully covered. Weeds can therefore germinate over a long period. Allium crops are also very sensitive to weed competition, so weed control is of particular importance. Weeds are more of a problem in crops grown from seed than in transplanted or vegetatively propagated crops. In general a drilled onion crop will produce no yield if it is left unweeded. At crop harvest, weeds foul undercutting and lifting machinery and prevent onion bulbs drying in the windrow.

The basic strategy for weed control in Allium crops starts with the choice of the field. The structure of the soil is important, and so is the preceding crop. The preparation of the seedbed is also a very important part of any weed control strategy. The seedbed should be fine, crumbly and level. Allium crops should not be sown or planted too early, otherwise the crop will grow slowly, giving the weeds the advantage of a longer germination period. Because perennial weeds are very difficult to control in Allium crops, they have to be controlled in the preceding crop.

Crop establishment generally follows conventional primary and secondary cultivations. The main methods of weed control are mechanical and thermal. Mechanical control includes harrowing and hoeing, while thermal control involves flame weeding to control small seedling weeds. The success of these methods depends on timing, on weather and soil conditions, and on the composition and density of the weed population. Crop rotation is important for disease control but volunteer weeds can be a problem when following crops such as potatoes, cereals and oil-seed rape.

Artichokes

Jerusalem Artichokes: land should be cleared of perennial weeds in previous rotations. The crop is fast growing and able to smother annual weeds. The artichokes themselves are liable to become volunteer weeds in following crops if not all the tubers are harvested. The plant is capable of rapid spread by means of rhizomes and tubers although tubers and rhizomes that fail to shoot do not persist in the soil for more than a year.

Asparagus

It is vital that the land is free from perennial weeds before the crop is established. Cultivation is not possible within the row during spear production. Soil cultivation to build up and maintain the planting ridges allows for some weed control. Ridging may be done by hand but mechanisation is used on larger plantings. In early autumn the fern is cut down and cleared and the soil on the ridges and in the valley bottoms lightly cultivated to improve the tilth for ridging in late autumn. Shallow cultivations may be possible to control winter annual weeds that have emerged. In spring cultivations are kept to a minimum on the ridges to avoid spear damage. The in-row weeds may be controlled by hand-weeding, mulching or with the use of geese. Flame weeding has been used against annual broad-leaved weeds but little information is available on crop stage. A low-growing living mulch that will suppress weeds is another possibility but there may be some competition with the crop.

Beans

Beans such as field beans and runner beans can be an important component of both arable and vegetable rotations. Weed management will obviously depend on the specific situation in which crops are being grown and the ultimate market. Weeds are a particular problem in crops grown for processing. At crop harvest, easily uprooted weeds can contaminate the crop with soil and stones carried on their roots. Other weeds become entangled in the harvesting machinery. Berries, flower heads and seed capsules can contaminate the final product unless removed by flotation or by hand.

Field beans: Autumn-sown field beans are generally resilient to weed interference and in some field studies the maximum yield loss was only 33% although in studies of the time of weed emergence relative to the crop, the earlier the weeds emerged the greater the level of yield reduction.

The cultivations needed for seedbed preparation depend on the weeds present and the time of sowing. Disking may be required if there is a high weed population present. Later sowings demand a finer seedbed and may involve using spring-tines. There may also be time to include a stale seedbed approach. Broadcasting the seed or using narrow row spacing gives greater weed suppression but reduces the weed control options. Wide row spacing facilitates mechanical weed control. The number of passes depends on weed density and crop and weed growth stages. Weed control may be initially with spring-tine harrows then inter-row hoeing. Blind harrowing and inter-row cultivations can keep the weeds under control at early crop stages. Once the early weeds are controlled, field beans tend to smother out later emerging weeds.

Beets

Three cultural types of beets (Beta vulgaris) are commonly grown in the UK including sugar beet grown as a field crop, fodder beet grown as a field crop for its roots and foliage for animal consumption, and beetroot, red or table beet grown as a vegetable crop outdoors or under protected conditions for the edible roots. Weed management practice will depend on which beets are being grown.

Beetroot:the optimum time for weed removal is around 3-4 weeks after crop emergence. Once the weeds have been removed the crop has some capacity for recovery from a check to growth due to the weeds Land should be free of perennial weeds. Conventional primary and secondary cultivations are used to establish the crop. False or stale seedbeds will reduce weed numbers in the growing crop. As with sugar beet, once the crop has emerged, regular inter-row cultivations with brush weeders, ridgers, steerage hoes, finger-tines will deal with weeds between the rows. There will be some effect on intra-row weeds but hand weeding may be required.

Fodder beet: May be drilled or broadcast. Punch planting makes use of the stale seedbed technique but minimises soil disturbance even further by dropping the seed into holes made by a dibber. This technique has been shown to reduce weed density by 30% compared with a normal drilled crop. Pre-emergence flaming has been shown to reduce weed numbers by 34 to 44 %. However, in relatively low value arable crops such as fodder beet that are grown on a large scale the cost of flame weeding may not be justified.

Sugar beet: Sugar beet usually follows a grass/legume leys or a cereal in the rotation and precedes a cereal or some other crop that will benefit from the residues of any manure application. Sufficient nitrogen from manure or compost application is important to ensure rapid the leaf development that will provide a dense leaf canopy and shade out the weeds. The primary and secondary cultivations required for seedbed preparation will have a considerable influence on the weeds. However, the nature and timing of these cultivations will vary with the previous crop, with soil type and with soil condition at the time of any operation. In general a level crumbly seedbed will give the crop the best start but growers may want to keep cultivations to a minimum. Weeds tend to emerge better and in greater numbers from a fine seedbed than a coarse one, but control measures are often more effective on a fine level seedbed.

It is important to achieve a good crop stand, as it is the dense leaf canopy that shades out emerged weeds and inhibits later flushes of seedling weeds. As the crop canopy does not close fully until mid-summer and tall growing weeds such as fat-hen and certain mayweeds may grow above the canopy before it closes. In the UK, the optimum weeding period is between 4 and 6 weeks after 50% crop emergence. In practice, weeding operations should commence at the 4-6 leaf stage and may cease at around the 10-12 leaf stage. Once the optimum weeding time has been reached yield may be depressed by 1.5% for each day the crop is left unweeded, although sugar beet has some ability to recover from an early check.

Brassicas

Brassicas can follow on directly after a ley or follow a cereal crop. The land should be ploughed and prepared early in the year to allow time for cultivation before planting that will kill any emerged weeds. Following crop establishment, weeds should be controlled without delay. Crop cultivars vary in the ability to suppress weeds because of differences in morphology. In vegetable brassicas the choice of cultivar may be limited by the need to schedule maturity to achieve continuity of production. The use of crop covers to advance maturity and protect vegetable brassicas from insect pests will hinder weed control and may enhance weed emergence and growth. Precise row spacing and careful alignment of cultivating tools facilitate mechanical weed control. Many vegetable crops, including brassicas, are grown on the bed system and cultivators and other implements should be matched to the bed widths. Weeding is repeated as necessary, and cultivations can become more thorough as the crop develops. When the crop plants are large enough not to be buried, tools can be arranged to move a 1-inch layer of soil towards and into the crop row. Where mechanical control is not possible, hand weeding may be required.

Broccoli: Transplanted broccoli rapidly develops a broad, shading leaf canopy. Inter-row weed control with a row crop cultivator, spider gang tool or a brush hoe 15 to 25 days after planting has given good weed control. Flexi-tine harrows provide adequate within and between row weed control but damage poorly established crop plants and reduce yield. Mechanical weed control with round and flat flexi-tines, brush hoe, rolling cultivator or shovel cultivator, alone or in sequence with flexi-tine weeders just prior to weed emergence at 13-26 days after crop planting has been shown to control weeds. As with cabbage, the broccoli crop has some tolerance to flame weeding and treatment at 2 weeks after crop planting has been evaluated.

Brussels Sprouts: Are generally a weed-suppressing crop and can be ridged up to improve weed control around the sprout plant. The wide spacing of the crop gives ample scope for mechanical weed control. Transplanted Brussels sprouts have an initial advantage over emerging weed seedlings, nevertheless, if left unweeded sprout yield is likely to be reduced by 13 to 24%. Close crop spacing (45 x 45 cm) increases weed suppression compared with wider spacing (61 x 61 cm) but within-crop competition then tends to reduce the yield of individual sprout plants. At closer row spacing, access for tractor steerage hoes becomes more difficult after early August. Early weed control is vital to avoid yield loss. As with cabbage, the Brussels sprout has some tolerance to flame weeding and treatment at 3 weeks after crop planting has been tried successfully.

Cabbage: Different types of cabbage include spring, summer, autumn, and winter cabbage including winter white and Savoy. Crops are established following conventional primary and secondary cultivations. Cabbage may be direct-sown but is more likely to be grown from bare-root or module-raised transplants. Good weed control is essential to maximise crop uniformity, quality and yield. Weed control is helped by an appropriate rotation that will avoid the build up of large weed populations. Making the most of weed control opportunities in one crop can frequently ease weed control problems in the crop that follows. The key period of weed control is the first four weeks after transplanting. Weeds that emerge after this are less likely to compete with a well established crop. In the drilled crop, single weeding at 3 weeks after crop emergence can give yields similar to those of the weed-free crop. In transplanted cabbage a weeding a single but thorough weeding 3 to 8 weeks after planting has been sufficient to prevent yield loss.

The preparation of a false or stale seedbed in the direct-sown crop may be used to reduce weed numbers before drilling the crop. The use of a false seedbed to reduce weed numbers in the transplanted crop is also feasible. Secondary cultivations may be used to kill emerged and germinating weed seedlings but must be shallow enough to avoid stimulating a further flush of weed emergence. Flame weeding will also kill the emerged seedlings and without disturbing the soil. The period from drilling to crop emergence is usually too short to provide an opportunity to kill early emerging weeds before the crop appears. If conditions allow, delayed drilling of the crop may give weeds more time to emerge and be killed before crop emergence. Careful timing of flame weeding is needed to kill the maximum number of weeds without damaging the emerging crop seedlings.

The crop should be kept free of weeds by surface cultivations using a tractor-mounted cultivator between the rows and hand hoeing around the plants. Soil should be drawn up towards the plants. Weeds between crop rows can be removed mechanically using a steerage hoe or brush-weeder. Both these machines work well but the brush-weeder has more flexibility to work in wetter conditions. The main advantages of the steerage hoe are the work rate, which is about three times faster than the brush-weeder. The main disadvantage with mechanical weed control is that weeds within the crop row cannot be removed, but plant spacing can be adjusted to allow earlier ground cover within the row to smother weeds.

Hand-hoeing between the rows may be used during early crop establishment, but the crop rapidly forms a dense leaf canopy that helps to suppress further weed development. Hand-labour requirements for weeding have been estimated at 50 man hours/ha. Inter-row cultivations with brush-weeders, finger-tine weeders, harrows or tractor-drawn hoes will control weeds within the crop row. The number of passes needed will depend on the weed population. As the crop develops the leaves spread into the inter-row and care is needed to minimise crop damage. Flexi-tine weeders with either flat or round tines used 14 + 24 days after transplanting gave adequate weed control and damaged the cabbage only slightly. Similar results were given by flexitime cultivations at 10 + 20 days followed by an S-tine row crop cultivator at 30 days after planting. The tines caused only slight damage to the cabbage.

Transplanted cabbage has a relatively high tolerance to heat, enabling band-flaming to be used along the crop row. However, the level of tolerance within each individual crop will depend on the growth stage of the crop and the waxiness of the leaves. After crop establishment, thermal weed control may be tolerated by cabbage if shields are used and the burners are directed away from the crop row. The aim is to kill seedling weeds in the inter-row with minimal soil disturbance. At an early stage most broad-leaved weed seedlings will be killed but grass weeds have a basal growing point and therefore a higher tolerance of flame weeding. Perennial weeds are also unlikely to suffer any long-term damage. The tolerance of most weeds increases with age and growth stage.

Mulches laid directly onto the soil surface provide a physical barrier to weeds. Materials available for use as mulches include; black polythene, black non-woven polypropylene and paper. Black polythene is the cheapest but water penetration is poor and drip irrigation systems may need to be laid first. Machines are available that can lay the mulch and plant through in the same operation. Black polyethylene is generally left down for the duration of a crop but studies have been made where the sheeting has been laid on the seedbed for much shorter periods and then lifted before planting brassicas. The short term covering of the soil with black polyethylene reduces subsequent weed emergence giving the crop an advantage over the weeds. Straw mulches have been evaluated in trials, but there appears to be a detrimental effect on yield, probably due to a temporary shortage of nitrogen as the straw decomposes.

Cauliflower: Should be placed early in the rotation to take advantage of higher fertility after the grass/clover leys. The seedbed should be weed free at the time of planting. One or two inter-row passes with a brush weeder, steerage hoe or finger weeder should be sufficient unless weed pressure or weather conditions result in poor control. Crop canopy closure should suppress later weed emergence and growth. As with cabbage, the cauliflower crop has some tolerance to flame weeding and treatment at 2 weeks after crop planting has been evaluated.

Kale: In transplanted kale, a single cultivation 10 days after planting was sufficient to prevent yield losses due to weeds. A further cultivation 10 days later reduced weed biomass but did not improve crop yield. Inter-seeding with the winter cover crops winter rye (Secale cereale), hairy vetch (Vicia villosa) or a mixture of the two after the final cultivation allowed the cover crop sufficient time to establish without sacrificing crop yield. Intercropping maize with kale for silage production can have advantages in terms of yield and weed suppression over the crops grown alone.

Swede: The crop is often sown late to avoid turnip fly and mildew, leaving time for cleaning the land before sowing. Inter-row cultivations can be used after crop emergence. The optimum time for weed removal is 6 weeks after sowing or 2-4 weeks after crop emergence. Weeds that emerged and competed with the swedes for up to 28 days after crop sowing had no effect if the weeds were removed and the crop then kept weeded. Swedes had some ability to recover after removal of weed competition.

Turnip: Has been considered a cleaning crop in the rotation because regular inter-row cultivations can be carried out until canopy closure. Sowing time is generally mid to end of May but wet conditions may delay this until June. The crop is often sown late to avoid turnip fly and mildew, leaving time for cleaning the land before sowing. Inter-row cultivations can be used after crop emergence. By sowing turnips after a ley all the weeds that germinated could be destroyed by cultivations. The optimum time for weed removal is 2-4 weeks after crop emergence. In the past, the inter-rows were horse-hoed and the rows were hoed when the crop was singled. A second weeding was carried out later if required.

Carrots

In drilled carrots left weedy through to harvest yield losses can range from 0 to 100%. Even low numbers of weeds can have a serious effect and weed biomass gives a better measure of potential crop losses than weed density. They have little ability to recover after removal of weed competition. Keeping the crop weed-free for a period of up to 42 days after sowing is very effective in preventing yield loss. Hand weeding may be cost effective in organic carrots but growers may spend 100 to 300 man hours per ha on hand weeding. Stale seedbed cultivations, flame weeding or brush weeding and other inter-row cultivations can reduce the need for this and are more economic in the long term. On land with a low weed population, pre-emergence flaming followed by a single weeding at 3 to 5 weeks after crop emergence has been sufficient to prevent crop losses due to weeds. For post-emergence weed control, a steerage hoe was more successful than a flame or a brush weeder in the situation under which the trial was done. The weed flora was relatively tolerant of flaming and a hard soil cap made the brush weeder ineffective. Covers (applied for early emergence or carrot fly control) also increase weed growth up to four-fold and additional weeding may be needed under protected crops.

Celery

Celery intercropped with leeks has been shown to aid weed suppression and increased the overall yield from the cropped area but reduced the quality of the leeks.

Lettuce

In drilled lettuce left weedy through to harvest yield losses can range from 0 to 100%. However, because of the effect on crop quality marketable yield is often zero.

Any site should be free of perennial weeds. The seedbed should have a fine tilth and if prepared early enough there will be an opportunity for a weed strike and final shallow cultivation to reduce the weed population before planting. The stale seedbed technique can be effective if there is sufficient moisture for weed germination. Flame weeding applied post weed emergence and before lettuce planting was most successful when weeds were small.

Precise row spacing and careful alignment of cultivating tools facilitate mechanical weed control. Closer spacing of lettuce may suppress weeds better but it will make mechanical or even hand weeding difficult should it become necessary. A minimum row spacing of 25 cm is needed to allow mechanical weeding. One or more passes with a brush weeder, steerage hoe or tine weeder may be required depending on the weed population. All can be equally effective but certain implements are better than others in particular conditions. A first pass may be needed 3-4 weeks after planting. Weeds should be dealt with while small. Hand-labour is not normally needed but has been estimated at 20-25 man hrs/ha. Tractor hoeing will take 5-6 hrs/ha.

Fleece covers are sometimes used for early production and for pest control. Weeds also benefit from the conditions under the covers, emerging in greater numbers and growing faster.

Parsnip

Parsnips are slow to emerge and slow to develop therefore early weed control is very important. As with carrots, the use of stale seedbeds and pre-emergence flaming followed later by crop ridging can aid weed control.

Pea

The effect of weeds on the yield of peas depends on relative times of crop and weed emergence and differences in seasonal rainfall. Peas that emerge first generally suffer less competition. In vining peas, increasing crop density reduced the total fresh weight of the weed flora but at higher crop densities, within crop competition increases and can affect yield so this has to be balanced against the benefits of weed suppression. Even at high crop densities weeds can still reduce yield through competition while at low crop densities the weeds reduce tillering of the peas. Weeds also reduce vining throughput during crop harvest.

Cultivar choice and seeding rate are important aspects of weed control in organic pea production. Seeding rate should be as high as economically possible, 120 plants per m-2 being seen as the minimum. Vigorous cultivars that grow fast and have a high biomass accumulation are needed to achieve the greatest competitive ability. Larger-seeded cultivars tend to be more competitive than the smaller petit pois types. Zelda and Ambassador are good quality high yielding varieties.

The semi-leafless nature of many modern cultivars allows greater light penetration through the crop to the weeds. In tests of the competitive ability of pea cultivars it was shown that vigorous types had a higher competitive ability. Smaller cultivars and semi-leafless peas appeared less competitive.

Chain harrows can be used pre-emergence and even post-emergence of the crop if compensated for by higher seed rates. In vining peas a single, relatively late inter-row hoeing when peas had 4-9 nodes and weeds were 5-10 cm tall, generally controlled weeds better than an earlier treatment. A sequence of an early followed by a later cultivation did not given any better weed control. Additional treatment with a torsion weeder gave only slightly improved weed control.

Radish

In favourable conditions, radish emerges earlier and grows faster than naturally occurring weeds. The crop is able to mature before the onset of weed competition. In addition, few weeds have reached the flowering stage before crop harvest so seed shed is not a problem. There is therefore no need for a carefully timed weeding if crop yield is the only concern.

Squashes

Marrows and Courgettes: There is usually the opportunity to prepare a stale seedbed and then use inter-row cultivations and hand-weeding to deal with subsequent weeds. Marrows and courgettes may also be planted into a black polyethylene mulch. The system is expensive due to the costs of materials and labour. The plastic mulch could be combined with the use of crop covers for earlier harvesting.

Sweet Corn

Weed competition trials with rows 25, 51 or 89 cm apart (with plants 30 cm apart in the row) demonstrated that in unweeded plots there was somewhat less yield loss in the narrower crop rows. Keeping the crop weed-free early, for 2, 3, 4 or 5 weeks was better than leaving the crop weedy for 5 weeks and weeding from then onwards.

Trials on inter-row cultivation with a row crop cultivator, a spider gang tool or a brush hoe did not provide adequate weed control in sweet corn, nor did flexi-tine harrowing. Cultivations must be shallow to avoid root damage. Flame weeding has given short-term weed control but could not maintain control of germinating weeds through the season. Others suggest corn can be flame weeded when it reaches a height of 4 inches and can be flamed until canopy closure. If flamed earlier than 4 inches tall the crop should not be treated again until it reaches a height of 6-8 inches. The crop row is flamed across with burners mounted in pairs, but staggered to avoid overlap, and set at an angle of 30 to 60 degrees from the horizontal. In addition to directed burners the use of leaf protectors reduces crop injury.

Fruit

Soft Fruit Crops

Weeds hamper picking and make ripening uneven. Newly planted crops are particularly sensitive to weed competition.It is vital that soft fruit crops are given the best growing conditions possible to give them an advantage over weeds. Soil preparation should be thorough to allow crop roots to penetrate as deeply as possible. Sub-soiling to break up the soil pan is an important preparation. Before crop planting the soil should be moist enough to sustain the crop. Bare roots should not be allowed to dry out or the crop will suffer a check to growth that will reduce yield.

Strawberries: As in other row crops, strawberry normally has a limited ability to cover the ground to suppress weeds. This is most evident in the year of transplanting when annual weeds may be problematic. In established strawberry fields, the cover of strawberry plants is normally good enough to suppress the annual weeds to some extent in the row. Perennial weeds may be problematic even within the rows. Between the rows both annual and perennial weeds have no competition from the crop. Annual weeds germinating in spring-planted strawberries had no effect on crop growth if removed by late May. Delaying weed removal inhibits stolon growth and leaf production but crop plants are able to survive the competition. Allowing weeds to remain until July, August or November in the planting year, reduces crop yield in the following year by 34, 54 and 67% respectively. It is important that the main spring flush of weeds is controlled. Under some circumstances weeds germinating in June may also adversely affect crop growth.

The land where strawberries are planted should as far as possible be free from perennial weeds. A traditional fallow based on cultivation can be an effective pre-planting treatment against shallow-rooted perennial weeds, such as common couch (Elymus repens), but to a lesser extent against deep-rooted broadleaved weeds such as creeping thistle (Cirsium arvense) and field bindweed (Convolvulus arvensis) . A bastard fallow may be used to clean up land containing some perennial weeds but strawberries should not be grown where creeping thistle is a serious problem. The planting bed can be prepared in advance using the stale seedbed technique allowing weeds to be flamed or rotovated before crop planting. Mechanical weed control is possible between the rows where the strawberries are planted in single rows without polythene mulch. A range of cultivators is used including brush weeders, finger weeders and rotary or rolling cultivators. Within the rows, straw can help to suppress weeds. Straw is also used between the beds. Perennial weeds may need to be removed by hand.

Where strawberries are planted into polyethylene mulch, the sheeting can be laid up to 4 weeks in advance of planting to encourage weed seedlings to emerge and die under the covering. Sheeting of 38 microns thickness is used for 1-year crops and 50 microns thickness for 3-year crops. Woven plastic mulch (Mypex) will last 9-10 years, much longer than black polyethylene, but is expensive. Straw mulch, 15 cm deep can be laid in the pathways to keep down the weeds. Straw can also be used as an alternative to plastic mulch.

Rhubarb: Covers the ground early and for much of the season which helps to suppress annual weeds but the land must be clear of perennial weeds. After planting on clean land the rhubarb should be mulched heavily thereafter. In one set of trials with rhubarb a straw mulch 15 cm thick controlled weeds better and was more cost effective than herbicide or hand-weeding treatments. The straw-mulched plots produced larger plants and higher yields in field trials over 6 years. A layer of straw applied before January protects young growth from frost and brings the crop on earlier. In Yorkshire, wool shoddy is still applied as a nitrogen source to rhubarb roots being grown for the 2 years prior to forcing. Weeds known as the wool aliens have been introduced in this way over many years. The country of origin of the weeds depends on the source of the wool.

Bush Fruits: Plantations, once established can remain productive for many years. New plantations should be laid out on land that has not grown fruit crops previously. Bush fruits can follow arable crops or grassland. The land may require subsoiling to remove compaction or shallow cultivation to break up the vegetation cover prior to ploughing. Soil inversion will bury established weeds and encourage further seed germination. If the soil is left for a while the seedling weeds can be killed by flaming or shallow cultivation. Perennial weeds may require removal by hand but land infested with perennial weeds should be avoided as far as possible. Black polyethylene entire or woven sheeting is widely used to suppress weeds within the crop row. A range of other materials is also used including straw, bark chippings, old carpet, grass clipping and composted organic materials. Some organic materials are likely to lock up nitrogen as they decompose. Loose materials need to be applied as a substantial layer, at least 3 inches deep to prevent weed seedling emergence. Where polyethylene mulch is going to be used, a trickle irrigation system may need to be put in place before the plastic sheeting is laid if irrigation is likely to be required.

Top Fruit Crops

Most orchards now have grassed alleys but recently established crops are particularly sensitive to competition. Even the grass strips can reduce extension growth by 20 to 60% if too close to the crop row. Repeated inter-row cultivations may be needed initially supplement with hand-hoeing. In widely spaced crops the area between rows could be ploughed or rotary cultivated. Cultivations will bury the emerged weeds but may stimulated further flushes of emergence and spread perennial weeds. Once the crop is established, soil disturbance may damage the surface roots.

Apple: Weed control by mulching with entire or woven polythene sheeting is possible but is difficult to lay around established trees. Once laid, the mulch needs to be cleared off periodically to prevent weeds establishing on top of it. In a comparison of mulches of asphalt paper, bark, plastic, barley- and rape-straw, with flaming and mechanical weed control, covering with straw gave the highest yield. Yield in plots covered with bark was low because common couch grew through the 10 cm layer of mulch. Flame weeding did not control annual meadow-grass.

Miscellaneous Crops

Grassland

Grassland or clover/grass leys are an important part of the organic farming system in the UK. Up to 70% of the farmed area comprises of mixed grasses and legume leys. Grass may be managed as a short, medium or long-term crop and this may determine the composition of the desirable sward species and the nature of the associated weeds. A survey of 502 farms in England and Wales where grass was the major crop (and at least half was permanent grass) found that 50% of farmers thought thistles, chiefly creeping thistle, (Cirsium arvense) were a problem while 40% considered docks (Rumex spp.) to be a problem. Thistles were mentioned more by beef farmers while docks were highlighted by dairy farmers. Docks appeared to be associated with low potassium and high phosphate while the opposite was true for thistles.

Oil and Fibre Crops

Oil Seed Rape: In organic crops broad-leaved weeds are considered to be less of a problem than wild oats and blackgrass. In a survey of conventional winter oilseed rape in central southern England in 1985 cleavers was the most frequent weed being found in 57% of fields whilst common poppy, prickly sowthistle (Sonchus asper) and scentless mayweed (Tripleurospermum inodorum) were also common (21, 18 and 14% of fields respectively). Volunteer barley can have a severe effect in the autumn on the growth of winter oilseed rape and, in Canada, it has been shown that increasing the seed rate improves crop yield when volunteer barley is a problem but there is no advantage when the crop is weed-free.

Linseed/Flax: The crop is not competitive and cannot compete against common chickweed, charlock, wild oats and other quick growing annuals. Fields should be free of perennial weeds. Experiments to determine the competitive effects of weeds on spring-sown linseed found wild oat (Avena fatua) and knotgrass (Polygonum aviculare) to be highly competitive. Chickweed (Stellaria media) and fat-hen (Chenopodium album) were less damaging. Chickweed formed a low mat that did not affect the crop growth while the fat-hen did not become as vigorous as it does in many crops. In winter linseed, chickweed and cleavers could be left in the crop until March-April without seriously jeopardising yields. Flax cultivars differ in their susceptibility to the allelochemicals produced by wild oats and some other weeds.

Hemp: There is time before drilling in late April to prepare a stale seedbed. Once the crop is established the foliage is dense and suppresses weeds, minimising further weed control needs. Weed free crop material is essential for processing.

Potato Crops

Weeds reduce crop yield by an average of around 36% but losses can be anything from

14 to 80%. Weeds are not always a problem in potato but control may be considered necessary to safeguard crop quality and yield. Perennial broad-leaved weeds including creeping thistle, field bindweed, the docks, and the perennial grass weeds, common couch and black bent are particular problems in potatoes. Among the annual weeds, taller species such as fat-hen are the most problematic. In the UK a single weeding between 2 and 8 weeks after crop planting has been shown to be sufficient to prevent a significant yield loss.

In potato cultivars, greater competitive ability is associated with early emergence, rapid early growth and the ability to develop a dense leaf canopy. Chitting the tubers aids early establishment of potato crops. Cultivars may differ both in growth rate and habit which may affect the time to canopy closure. Some cultivars are known to produce taller and leafier foliage than others. However, mechanical weed control can damage lush foliage especially later in the season. Increasing planting density may improve weed suppression but it also increases establishment costs and may encourage blight due to the greater density of foliage.

The normal cultural practice is to ridge shortly after planting and let the ridges settle. Weed control is then applied ten days after planting using chain harrows, ridgers or purpose built weeders. Harrowing down the ridges between planting and emergence, hastens crop emergence as well as controlling weeds. The number of passes depends on weed density. Where a second harrowing is needed this may be carried out at or shortly before crop emergence. Thermal weed control can also be used to control seedling weeds prior to crop emergence.

Inter-row cultivations between the ridges and re-ridging are carried out as needed post crop emergence. Cultivations are best done when weeds are small and unlikely to re-establish. Rolling cultivators used by some growers have tines that weed between the rows and rolling star-shaped tines to cultivate the sides of the ridges. The ridges are then rebuilt by ridging bodies that follow the tines. In addition to controlling weeds, earthing up the ridges covers the developing tubers and reduces the incidence of tuber blight. Post planting cultivations should be kept to a minimum to limit water loss and crop injury. Crop damage can result in an average yield loss of 7% but losses due to weeds can be considerably higher.

There has been some interest in the use of living mulches for weed control. Cover crops such as vetch, oats, barley and red clover have been evaluated for their potential ability to suppress early-season weeds in potatoes where the cover-crops were inter-seeded at ridging or 3, 4 or 5 weeks after crop planting. Weed control with red clover was consistently poor. The cereals and vetches gave good weed suppression but reduced crop yield. The application of green manure material from mustard and oilseed rape can result in weed suppression due to the release of allelochemicals as the mulch decomposes. In the USA, an autumn-sown rapeseed green manure crop incorporated in spring prior to potato planting reduced weed density by 73 and 85% in different years, and reduced weed biomass by 50 and 96% in those years. Mulches of paper, plastic and other materials have given good control of weeds but are economic only in high value early potato crops. Where early potatoes are grown under floating covers, weed control is difficult until the covers are removed.

Disease Management in Organic Crop Production

The idea of "disease control" is somewhat misleading. It suggests we can find a solution to a disease and then the problem will no longer exist. The reality is that any recommendations can only reduce losses caused by disease. They do not eliminate disease altogether.

The term "disease management" more accurately conveys the impression of an ongoing process to reduce disease to acceptable levels. Disease management involves making conscious decisions related to numerous agronomic factors over which growers have some controls.

Plant diseases are caused by microorganisms such as fungi, bacteria, viruses and nematodes. A simple model called the disease triangle describes the conditions that govern disease severity.

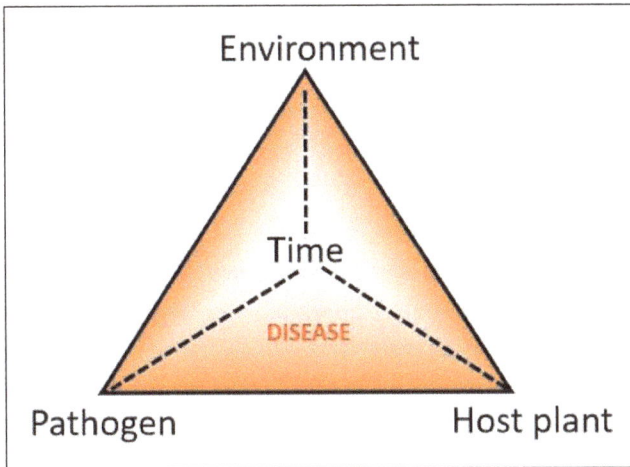

The Disease Triangle.

In order for disease to develop, three factors must be present: a suitable host plant, an infectious pathogen (microorganism that causes the disease), and a suitable environment. Disease control measures aim to reduce or eliminate one of the corners of the disease triangle. Since the three factors (host, pathogen and environment) interact, it's important to consider each one and incorporate a variety of management approaches.

Crop Rotation and Intercropping

Rotation of susceptible and resistant crops is one of the oldest practices used to control disease. It remains an important practice against many diseases, where a specific control, such as host resistance, is not available. Rotation is particularly effective in controlling soil- and stubble-borne diseases. The success of rotation in disease reduction depends upon many factors, which include the ability of a pathogen to survive in the absence of its host and the host range of the pathogen. Those that have a wide range of hosts will be controlled less successfully. For example, sclerotinia stem rot of canola (Sclerotinia sclerotiorum) has been recorded on about 400 different plant species worldwide. In Saskatchewan, it is known to occur on 13 field crops. Pathogens that live indefinitely in the soil are less likely to be curtailed by rotation than those that can survive only brief periods apart from their hosts.

Transmission of pathogens via seed, the presence of susceptible volunteer crops and weeds that harbour the pathogens, and the distribution of pathogens by wind and other means will reduce the benefit derived from a crop rotation. For instance, rotations are ineffective in controlling rusts in small grain cereals because the rust fungi do not overwinter in western Canada. Inoculum from the south is disseminated by wind in summer. In a similar fashion, diseases such as barley yellow dwarf and aster yellows depend largely on the northward movement of insect vectors, often over several hundred kilometres.

Crop rotation should be used in conjunction with other cultural practices to maximize its benefit. In establishing a rotation, select crops that are as diverse as possible. In general, a disease may infect closely related species, but will not injure those that are unrelated to the natural host (e.g. net blotch in barley).

Intercropping is growing two or more crops together in the same place at the same time. The original goal is to find synergies between the mixed crops or ways in which the plants do better together than they do alone as single crops.

Intercrops can change the environment for the stand. For instance chickpea is vulnerable to ascochyta blight. Intercropping chickpea with flax alters the crop canopy and improves airflow. This reduces the spread of disease in chickpea. Similarly improved airflow movement in intercrops of pea and oats can reduce pea disease and an intercrop of heritage and modern wheat cultivars reduces the overall disease pressure in the stand and stabilizes yields.

Cereal Grains

Crop rotation is recommended to control a number of leaf blights of small grain cereals. Leaf diseases such as net blotch (Pyrenophora teres), scald (Rhynchosporium secalis), and speckled leaf blotch (Septoria passerinii) of barley, and tan spot (Pyrenophora tritici-repentis) and septoria blotches (Septoria spp.) of wheat are generally not carried long distances by wind. The pathogens overwinter in crop debris and on the seed. A break of two to three years between susceptible crops will markedly reduce or control these diseases. This rotation break will allow previously infected crop residue to decay in the soil. Sanitation to eliminate volunteer host plants and the use of clean seed is important.

Soil-borne diseases such as take-all (Gaeumannomyces graminis) may be controlled when wheat follows non-susceptible crops, such as oat, pea, flax, sweetclover, sunflower, or fallow. Barley is less susceptible than wheat, but is still affected by the disease. Including barley as a break crop in the rotation should not be considered as it maintains the disease for future susceptible wheat crops. Unlike take-all, common root rot of cereals is caused by several fungi which are both seed and soil-borne. The control of this disease is more difficult to achieve through rotation, particularly in areas where mainly cereals are grown. Non-host crops for common root rot include canola, flax and legumes. However, lentil can be infected by a specific rot root organism (Fusarium avenaceum) which also infects cereal crops.

Ergot occurs on a wide range of cereal plants and grasses. It is most prevalent on rye and triticale but also attacks durum wheat, common wheat and barley, in decreasing order of susceptibility. A rotation allowing one year between successive crops of rye, triticale, wheat or barley will significantly reduce or control this disease. If native or forage grasses border a cereal crop, mowing the grasses at heading will reduce ergot in the adjacent cereal crop.

Oilseed Crops

Canola acreage is now limited in organic agriculture; however, the two major diseases that affect this crop are blackleg (Leptosphaeria maculans) and sclerotinia stem rot. For blackleg control, a rotation to non-susceptible crops for a period of three to four years is vital. Blackleg survives in canola stubble which generally takes a long time to decompose. Mustards (brown, oriental and yellow), cereals, and legume crops are resistant to this disease and can be safely used in a rotation.

At one time, sclerotinia was thought to be controlled moderately well by crop rotation. Research has indicated that little difference exists among crops in two, three or four year rotations. Rotations of four years or longer may offer some control. This length of rotation may not be practical in canola growing areas. Sclerotinia also attacks pulse crops, sunflower, mustard, sweetclover and flax. This further limits the crops that can be used in a rotation to manage the disease.

Other diseases such as white rust and staghead can be a problem on polish canola. Since both of these are stubble-borne, crop rotation is an effective method of control.

Pulse Crops

Ascochyta blight can infect numerous pulse crops such as field pea, lentil and fababean.

Each crop has a specific ascochyta blight fungus. Thus, one does not have to worry about ascochyta from one pulse crop infecting another in a rotation.

With lentil, follow at least a three year rotation (i.e. two year break) involving a cereal, oilseed or other pulse crop. This should be longer (up to four years) if the pulse crop residue is resistant to breakdown (e.g. fababean). Stubble residue is the main source of inoculum for this disease. In field pea, mycosphaerella blight (Mycosphaerella pinodes) can also be introduced as airborne inoculum.

Powdery mildew (Erysiphe polygoni) is specific to field pea. Crop rotation will help, depending on climatic conditions. Dew formation, and lack of rainfall, favours the development of the disease. Inoculum is spread by wind and, once established, powdery mildew increases very rapidly. Field isolation may assist in reducing infection which occurs by wind movement.

Root rot diseases can be a problem on pulse crops. Since the pathogens involved do not usually infect cereals, they can be reduced in severity by including a cereal crop or flax in a rotation. Unfortunately, control is complicated by the wide range of alternate non-crop host plants which these pathogens can infect.

Flax

Flax is a crop with a limited range of disease problems at present. Serious diseases such as rust and wilt are controlled by disease resistant cultivars. Pasmo (Septoria linicola), seed decay and seedling blights that are specific to flax can be controlled by using rotations with several years away from flax. Careful harvesting and handling to prevent seed coat cracking can reduce seed decay and seedling blight.

Forage Legumes

Forage legumes such as alfalfa, sweetclover and red clover have numerous disease problems that affect both leaf and root tissue. Almost all major diseases of forage crops are amenable to control through crop rotation. The wilts (verticillium and bacterial), crown and root rots and foliar diseases all survive on dead plant material. When a forage crop is plowed down and another crop is planted, there is a period after which the disease inoculum is reduced and dies out. This time period will vary, depending on the time needed for the forage residue to decompose.

Resistant Cultivars

The use of cultivars resistant to the prevalent diseases in the growing region is the most efficient and cost effective means of disease control. There is little additional operating expense and no hazard to the producer or the environment. Equally important is the fact that crop residue of resistant cultivars constitutes less of a disease inoculum problem for future crops.

The use of resistant cultivars in conjunction with other crop management factors, such as field sanitation and crop rotation, can prevent many serious diseases from occurring.

The use of resistant varieties is particularly beneficial where crop rotation is of limited use. This is the case when the disease organism has any of the following properties:

- The pathogen has a wide range of hosts (common root rots, seedling blights).

- The pathogen has a long lived resting stage or persists in the soil for a long period of time (fusarium wilt in flax).

- The pathogen has airborne spores (cereal grain rusts), or is carried by insects (yellow dwarf, aster yellows), that are spread over long distances.

- The pathogen has a very high rate of spread (powdery mildew).

- The pathogen is seed-borne (cereal smuts, ascochyta blight of lentil, pasmo).

- The use of the term "resistant" cultivar may be misleading. In the case of ascochyta blight in field pea, the difference in resistance ranges only from very poor to fair. With blackleg in canola, the level of "resistance" ranges from very poor to very good. Growing these "better" varieties does not mean that producers can forget about using rotations or any other management technique that will reduce the level of infection. Such variation in level of resistance within a crop, between varieties, occurs for a number of diseases.

Sanitation Field

The importance of managing disease-infected straw and chaff cannot be overemphasized. For all crops and numerous diseases of these crops, straw is the primary source of future inoculum. Incorporating this residue into the soil by tillage hastens the destruction of the pathogen by beneficial fungi and bacteria. In addition, diseased plant material which is buried under the soil surface prevents future movement of the spores by wind.

Equally important is the value of maintaining this residue on or near the soil surface for conservation and soil improvement. The balance between soil conservation and disease control will vary depending on geographical location, soil type and disease prevalence.

On lighter textured soils or in regions where soils are prone to water and wind erosion, stubble should be left standing over winter, with tillage postponed until spring. With spring-seeded crops, the incorporation of crop residues into the soil, if desirable or necessary, should be done just prior to seeding. Only in exceptional circumstances, such as alfalfa trash heavily infested with leaf pathogens, should stubble burning be considered.

Alternate disease hosts include volunteer plants growing within a crop or weeds along field margins and fence lines. They should be controlled. Left uncontrolled, these types of plants can transmit disease to healthy crop plants. In many cases, the unsatisfactory control of diseases within a rotation (break crop) may be attributed to inadequate control of volunteers.

In the case of ergot in cereal crops, tillage prior to seeding can have a useful function. Cultivation of the soil surface will bury the resting bodies (ergot), preventing them from germinating and infecting succeeding crops.

Seed

Seed should be uncontaminated with sclerotia, ergot bodies, smut or bunt balls and free of infected or infested kernels. Many pathogens are seed-borne, such as ascochyta blight in lentil, smuts, barley stripe mosaic (a virus), leaf stripe of barley, pasmo and numerous others. Seed supplies may be analyzed for various diseases by forwarding samples to a seed testing laboratory. It should be noted, however, that relatively few diseases are exclusively seed-borne. It is more common for pathogens to be transmitted in soil or on stubble as well as with the seed. Ascochyta blight of lentil is transmitted on seed and stubble. The beneficial effect of a good crop rotation may be lost or reduced if high inoculum levels are reintroduced into the field by contaminated seed. Seed quality is therefore important and one more factor to incorporate into an effective disease management program.

Sound seed is also important, particularly in crops such as flax, rye and pulses, where cracks in the seed coat may serve as entry points for soil-borne microorganisms that rot the seed after it is planted.

Planting Dates and Rates

Susceptible cultivars of some crops may escape significant damage from disease by planting at a time which avoids exposure to inoculum and thus severe infection. Diseases such as barley yellow dwarf can be controlled by early seeding. The aphids that transmit this disease do not arrive in significant numbers until early summer. Younger plants, which would have been seeded later, are more attractive as food to aphids.

Field peas are readily infected by powdery mildew. The severity of this disease can be reduced by planting early, as soon as soil can be prepared. If pasmo becomes a problem, early seeding, following the risk of frost, will help escape infection. If lentil seed is infected with ascochyta, planting should be delayed until soil temperatures are higher. This practice will reduce the disease severity.

For fall-seeded crops such as winter wheat, later seeding, following the maturation of spring cereals, breaks the life cycle of the mite which carries the wheat streak mosaic virus.

The density of plants and leaf area in a field is determined by seeding rate, plant architecture and growth habit of the crop, as well as climatic conditions. Dense foliage can increase the chance of leaf diseases by providing a larger surface leaf area for infection to occur. In addition a dense leaf canopy can create a moist soil surface, favourable to a pathogen such as Sclerotinia.

Reduced seeding rates may decrease take-all severity in spring wheat. Plant and leaf density and the ability to control it, as well as the disease potential because of it, should be considered. Note, however, that a reduction in the leaf canopy will create an opportunity for invasion by weeds. This may negatively affect overall crop productivity, depending on the particular situation.

Always consider optimum seeding depth. Deep seeding in cold soils can result in seedling blights and damping-off. This is particularly true in pulses or other small-seeded crops such as canola. Semi-dwarf cereals should not be sown deeper than four centimetres. Deeper seeding encourages root disease.

Soil Fertility

Soil Nutrient Levels

Soil fertility levels, or applications of fertilizers which create nutrient imbalances in soil, can have a marked influence on the crop's susceptibility to disease. Judicious applications in both amount and proportions of nutrients should be based on a sound soil testing program. Excessive levels of nitrogen can result in lush leaf tissue which creates an ideal environment for leaf pathogens or the vectors of viral diseases. In contrast, inadequate soil phosphorus can predispose wheat to browning root rot (Pythium spp.). Few diseases respond as dramatically to nutrition as take-all in cereals. Losses from this particular root disease are generally severe when plants are deficient in any of the essential elements.

Twelve of the 13 principal mineral nutrients are reported to affect take-all, either individually or collectively. A deficiency of nitrogen and/or phosphorus can result in a marked increase in take-all. The use of phosphorus in sufficient amounts to stimulate crop growth has a positive effect in reducing losses from take-all. A large single application of phosphorus may decrease take-all more effectively than small applications, providing other nutrients are not deficient and the phosphorous application is not so large that it inhibits the uptake of copper and zinc. The beneficial effects from phosphorus in reducing take-all are a reflection of stimulated root development and increased host resistance.

A shortage of micronutrients, such as zinc or copper, can cause "disease" in the form of deficiency symptoms in cereal, pulse and oilseed crops. The level of micronutrients required for sufficiency may be very low. However, deficiency in micronutrients can have serious consequences. An over-application of certain micronutrients, such as boron, can cause severe toxicity. Plant tissue analysis is the most accurate method for diagnosing problems of this nature.

Manure

The addition of livestock manure to improve soil and crop productivity also stimulates higher populations of soil microorganisms which compete with or destroy soil pathogens. Nutrients released from the decomposing residue may also stimulate the activity of some pathogens but, having no host to attack, they die. This positive process occurs with many root-invading pathogens of cereal, oilseed and pulse crops.

The soil incorporation of green manure cover crops such as certain cereal crops (barley and rye), field pea, sweetclover and lentil (cv. Indian Head) can be used effectively to reduce or control root rot pathogens and other diseases. Legumes, especially alfalfa following breaking, are very effective in suppressing disease organisms. Plant residues of this nature stimulate the growth of soil microorganisms which are antagonistic to plant pathogens, thus offering a measure of disease control.

With the addition of manure, it is possible to create an imbalance of nutrients in the soil. The quality of organic material and quantity of nutrients present in manure varies depending on the type, the method of storage prior to application, and even the application method. Following regular applications of manure, one should pay attention to monitoring the levels of nitrogen, phosphorus, sulphur and, in certain cases, micronutrients present in the soil. This should be done by soil testing on a regular basis.

Integration of Disease Management Practices

Disease organisms and their parasites, host crops, associated vegetation, soil and meteorological conditions are all elements of a linked interdependent system.

Since crop plants are subject to more than one disease, methods for the control of each disease must be woven together. In addition, most diseases are transmitted in more than one fashion. This will require integration of several control and management strategies at more than one level. Disease control strategies, in turn, need to be combined with methods for weed, insect and other production concerns. For example, clean cultivation, achieved by burying infected crop residue in the soil, is a highly effective way of controlling many diseases. However, it may be an unacceptable production practice because of its effect on soil erosion and water conservation. Attempts to increase yields by applying higher rates of nutrients, such as nitrogen, can lead to increased leaf disease problems. Manure incorporated into the soil can reduce the level of numerous pathogens. An over-application, however, may create nutrient or chemical imbalances resulting in micronutrient deficiencies.

Integration of disease control and crop production practices must be done carefully and suit the producer, location and type of cropping system. Disease management systems must be devised to utilize control technology that works harmoniously within the cropping system while maintaining the pathogen population below economically damaging levels.

Organic Crop Insect Management

Managing Insects in Organic Crops

Managing insects in the farm ecosystem is challenging. At a glance, it might seem that one insect is not significantly different from another, and that all are pests. But, because insect populations are interrelated, intervening to control them may prove to be more disruptive than beneficial.

Specific insect types must be considered when control strategies are devised. No single control method will be adequate for all. Successful management depends on incorporating a number of control strategies.

The ideal is to establish equilibrium between beneficial insects and pests and to use cropping practices that reduce insect populations. To reach that goal, consider the following management factors.

Crop Rotation

Planting fields with different crops effectively controls plant diseases and weed problems. It can also be used to reduce insect populations for: 1) species that feed on a single crop or related crop; 2) species where the overwintering stage is associated with that crop; and 3) species that have limited mobility.

Crop rotations reduce insect problems by denying easy access to the crop and increasing the abundance and effectiveness of insect predators and parasites. The diversified habitat provides these parasites and predators with alternative food sources, shelter and breeding sites.

Crop diversity is essential to limiting insect populations. A balanced cropping system, using a combination of cereal, oilseed, pulse, and biennial or perennial legume crops, helps limit insect populations. A diverse rotation will contain a larger number of insect species adapted to the various crops within the system. This in turn creates a proportionally higher number of natural enemies to any insect pest. Each crop type should be separated as far as possible within the rotation sequence.

Repeated cropping of large areas to single or closely related crops year after year encourages insects that feed on those crops to multiply rapidly and become pests. Single crop farming should be avoided.

Cabbage aphids, flea beetles, diamondback moth, cabbage butterfly, wheat midge, red turnip beetle, wheat stem maggot and wheat stem sawfly are a few of the many insect pests that can be regulated with diverse crop rotations.

Sanitation

Removing or destroying crop residue and alternate host sites can help control some insects. It is most effective against insects that overwinter in crop residue or lay eggs in it, and where crop residue protects insects from exposure to extremely cold temperatures.

Winter survival of wheat stem sawfly, Hessian fly, bertha armyworm, alfalfa plant bug and some cutworms increases when crop residues are left undisturbed.

Where wheat stem sawfly is a problem, shallow fall tillage is recommended. Tillage will also reduce bertha armyworm survival but may not be practical as a control strategy given the adult moth's flight range. Where tillage is required, attention should focus on maintaining a balance between crop sanitation and soil erosion protection. This will require leaving some crop residue on the soil surface.

In alfalfa seed fields, where cultivation is not practical, the alfalfa plant bug can be controlled by burning crop residues in late fall or early spring. Burning destroys the eggs laid in the alfalfa stems the previous summer and fall. In alfalfa produced for hay, burning is not required. Normal cultural practices, such as haying for forage, will remove the eggs and prevent a population build up.

Also attend to field margins and fence lines where weeds or volunteer crop growth occurs. The close proximity of volunteer plants could provide alternate habitats for insects to complete their life cycle until an appropriate crop occurs in the nearby field.

The size of a field can be important in determining the intensity of an insect problem. In general, it is a good idea to keep field sizes small.

Weed Management

Weeds should always be controlled to reduce their impact on crop yield and quality. Cultural control of insects may require a slight deviation from this concept. Sometimes a low, in-crop weed population may even be beneficial. In the case of aphids and leaf hoppers, their numbers on certain crops decrease as the diversity of host plants (weeds) increases. There are specific situations where this would not be true, such as for insects that are host specific.

The increase in plant diversity that results from allowing a certain amount of weed growth has been shown to have a positive effect on the diversity of insect species. The diversity may increase the number of predators attacking a pest, as has been observed in alfalfa stands. For example, an increase in plant diversity resulting from allowing a low population of weeds to grow increases the activity of parasitic wasps, which would help to control an outbreak of alfalfa caterpillar (Colias eurytheme).

Seeding Date

Planting should be timed to provide the crop with a competitive advantage over insects. Early seeding reduces crop damage caused by grasshoppers. It may also markedly reduce aphid damage on cereal crops. The crop grows and becomes vigorous and passes the susceptible stage before aphid populations reach damaging levels. Similarly, seeding wheat early may reduce wheat midge damage.

In the case of wireworm in cereal crops, avoid very early or late seeding. Problems with sweetclover weevil can be reduced by sowing clover early, before grain crops. Hessian fly problems in winter wheat can be avoided by seeding after mid-September, thus preventing egg laying in newly seeded crops.

Delayed seeding can also be an effective way of avoiding problems. Waiting until the insect has finished its life cycle before seeding or until the spring insect migration is complete can reduce crop loss.

Crop damage from army cutworm can usually be controlled by delayed planting. This allows the overwintering cutworms to finish their development before the crop emerges. However, a cool spring may slow larval development, extending the delayed seeding time interval beyond a practical limit.

Similarly, barley thrips can be controlled with delayed planting. In the absence of barley, the migrating thrips colonize the non-crop host plants instead of the barley.

Seeding Rate

Increasing plant densities using higher seeding rates can have a positive effect on insect pest control. By having more plants in the field, a given aphid population will have less of an impact upon individual plants.

In the case of aphids, leafhoppers and flea beetles in canola, the amount of bare ground present in a plant community appears to influence abundance. These insects are attracted to the contrast between a green host and a dark soil background. Increasing seeding rates obscures this contrast. Depending on environmental conditions, increasing the plant density could change plant geometry (less tillers and leaf area) and/or the "lushness" of the plant. Either of these factors could alter the attractiveness of the crop to insect pests.

Higher plant densities may result in certain insects being more abundant. Increased problems with the true armyworm have been reported in densely seeded stands of fall rye.

Summerfallow and Stubble Management

Depending on the geographic area, summerfallow can be an important aspect of a cropping rotation. It may be used for effective insect, weed or disease control. The management of summerfallow can have a large influence on certain insect populations.

It is important to keep volunteer crops and weed growth to a minimum, since these plants serve as a temporary host, defeating the purpose of summerfallow, which is intended to break the life cycle of many insects.

Destruction of green growth in summerfallow fields should begin early in the year before insects like grasshoppers begin to hatch. This will destroy their food and starve newly hatched grasshoppers. Early June tillage of stubble infested with wheat stem sawfly helps bury their larvae, increasing mortality rate. Shallow tillage in the fall will also increase mortality rates of this pest.

Severe infestations of the red turnip beetle in canola fields can be prevented by destroying their eggs with late fall or early spring tillage. Spring tillage to remove cruciferous weeds and volunteer canola will control the larvae, by eliminating food sources. Another sporadic pest of canola, the bertha armyworm, may be reduced in number by fall and spring cultivation. Pupae of this insect are thus exposed to freezing, diseases and predators.

Controlling winter annual and volunteer crops in the fall is critical. Several kinds of cutworms hatch in fall and overwinter as partly grown larvae. If available, these larvae will feed on weeds and volunteer crops until freeze-up and again in the spring. One species of the army cutworm sometimes causes damage to winter wheat and fall rye.

In years when it is abundant, this cutworm may also seriously damage spring-seeded crops in fields where sufficient green growth in the fall provided food for the young larvae. Redbacked and pale western cutworm moths also lay their eggs in the fall until mid-September, but eggs do not hatch until the following spring. Moths of the redbacked cutworm, which is primarily a pest in the parkland areas, usually concentrate their eggs in patches of weeds in summerfallow.

For the pale western cutworm, sequencing agronomic practices is important. It prefers to lay eggs in loose soil. Where this insect is a problem, summerfallow should be cultivated before the middle of August and left to crust over or cultivated after the middle of September. In the spring (May), a delay of five or more days between cultivation and seeding can prevent infestations. The larvae die if they feed after they hatch and then are deprived of food for several days or cannot feed at all for 10 to 14 days.

Summerfallow can be useful in cereal-only cropping systems to reduce the likelihood of insect damage, particularly to the first subsequent crop. This interrupts the build up of resident insect populations. Similarly, allowing a 10-day break between the harvest of a spring wheat crop and the emergence of a winter wheat crop is critical for the control of wheat curl mite. A time period less than this will allow the insect to transfer from one crop to the other, continue its life cycle, and infect the winter wheat with wheat streak mosaic.

Trap Strips

The function of seeding barrier strips of crop is to lure insect pests into a specific area. The pest problem can then be concentrated, making control easier. This approach can work for grasshoppers where damage is usually initiated from the edge of a field inwards.

Wheat stem sawfly may be lured into a temporary trap crop of a susceptible variety placed around the crop. This area is then swathed for hay or silage in July to remove some of the larvae population. A permanent trap crop of smooth brome grass around a field will reduce the number of larvae surviving in ditches and headlands.

Soil Fertility

Adequate, balanced soil nutrition is essential for crop quality, yield and moisture use efficiency. An imbalance in fertility can affect plant quality in several ways. Excess nitrogen, in relation to other nutrients, can result in plants becoming more succulent and thus more attractive to insects. Nitrogen is also responsible for stimulating vegetative growth, producing more leaf area. This would allow an overall higher pest infestation per field. At the same time, the enhanced growth rate resulting from this nutrient may offset potential insect damage. This depends on the particular growing conditions at the time of the pest problem.

A nutrient imbalance can readily occur where large applications of manure have been applied. Soil tests should be conducted regularly to monitor nutrient levels and allow corrections to be made.

Beyond a pre-established level necessary to maximize crop efficiency and yields, other factors become important.

Resistant Cultivars

Selection and use of resistant cultivars is a cost-effective form of insect control. For example wheat stem sawfly can be effectively controlled in the southern prairies by planting one of the "solid stem" wheat varieties. These cultivars are resistant to damage and should be used where crop yield is reduced by more than 10 percent. Other insects, such as the pea aphid, can be a pest on field pea, alfalfa and clover. Several pea varieties are not severely damaged by this aphid. In addition, certain varieties of alfalfa developed in the U.S. are resistant to the pea aphid.

Not all plants are equally attractive to insects. Factors such as leaf and stem toughness, pubescence (leaf hairs), nutrient content, water content and secondary compounds or toxins all influence an insect's growth or reproduction. Grasshopper development and reproduction is significantly better on wheat, followed by barley, which in turn are both better than oat. Similarly, grasshoppers usually won't feed on certain varieties of field pea (CV. Sirius). This insect will selectively feed on lentil pods, while causing minimal damage to stems. Many of these properties are used by plant breeders to develop resistant varieties.

Insect resistance can also occur simply as a function of plant architecture or geometry. Certain cereal crop varieties with pubescent leaves are less desirable to cereal leaf beetles. Within a species, differences in tillering, leaf shape and size and plant height all alter the environment available to the pest. Similarly, differences in maturity between varieties can be used to advantage, depending on the specific pest problem. Some factors may simply reduce the insect's developmental rate, reducing the problem to an acceptable level.

Intercrops

Intercropping is growing two or more crops together in the same place at the same time. The original goal is to find synergies between mixed crops or ways in which pants do better together than they do as single crops. The use of intercropping systems provides an option for insect control for organic farmer that are limited in their chemical use.

Intercropping can make it more difficult for pests to find their host plant. Many insects find their target by smell. The addition of a second crop in the field can disguise the scent for searching insects. Intercropping canola with barley reduces flea beetle and diamondback moth damage to canola. Similarly, intercropping mustard with barley or wheat reduces the problems with these insects. Another interesting example is midge resistant wheat. Midge resistant wheat is sold as a mixture with the non-resistant wheat in hopes of providing a refuge for the wheat midge and slowing the adaptation of the midge to the resistant wheat. in addition, intercropping systems for insect pest control can also include the planting of a crop that has a repellent effect, attractant effect or a combination of the two, on a targeted insect in close proximity to a crop that has the potential to be attacked by the insect.

Biological Control

In any balanced ecosystem, biological control by natural predators is constantly occurring. The more diverse a cropping system becomes, the greater the spectrum of insect species within it. This in turn leads to the development of more natural predators.

Biological control of wheat midge occurs from a parasitic wasp. This wasp can destroy about 40 percent of the overwintering population of wheat midge.

Ladybird beetles, ambush bugs, assassin bugs, lacewings and a host of other insects are predators of aphids, bertha armyworm larvae, sunflower beetles, etc. The pea aphid, which is a main pest of field pea, alfalfa and clovers, is reduced or controlled by damselflies, minute pirate bugs, ladybird beetles, lacewings, and hover fly larvae. In cereal crops, ladybird beetles can have a noticeable impact on aphid populations, usually keeping them in check.

Insects such as beet webworm are found in a wide range of weeds and crops, including canola, mustard, flax, sweetclover, alfalfa and sunflowers. They are attacked by a number of parasitic insects. Some of these parasites destroy the overwintering pupae. This is particularly true in lighter textured sandy soils.

The above examples of biological control occur naturally in a healthy, balanced system. We often think of biological control in terms of adding or "inoculating" an existing system with a foreign predator. The hope would be to gain pest control similar to the above examples. Microbial insecticides such as Bacillus thuringiensis (BT) would fall into this category. This bacterial pathogen is effective against a wide range of caterpillars (Lepidoptera). One great advantage of microbial pesticides is the specificity of their toxic action and their safety to non-target organisms. These products cause minimal environmental disruption.

Conservation of natural enemies refers to the practices designed to enhance the colonization and/or survival of natural enemies within the crop. Possibilities include providing pollen sources, nectary plants, shelter, or nesting sites. This approach is particularly well suited to perennial crops, where a long-term equilibrium between host and natural enemy can be established. This approach to pest control is relatively new and requires further research to be fully evaluated.

Insect Hormones

Synthetic insect hormones have been developed to duplicate naturally occurring insect (pest) chemicals. In nature, these products regulate insect body functions. Their mode of action is other than direct toxicity. When applied these products disrupt insect development, such as growth, rather than causing immediate death.

Depending on the type and amount, certain types of hormones (pheromones) may be used as attractants to monitor population levels or simply attract insects into a trap. These products can be categorized as having low mammalian toxicity, low use volume, target species specificity and natural occurrence. These products have potential, but require further development.

Organic Pest Management Approach

- Pests are indicators of how far a production system has strayed from the natural ecosystems it should imitate.

- Pests are attracted to a plant that is weak or inferior.

- In a well-balanced system, massive pest outbreaks are rare due to the presence of natural predators, parasites, and disease agents.

- Prophylactic, holistic approach vs. remedial approach.

- Not just treating symptoms.

- Pest problem usually indicates sub-optimal growing conditions and imbalance.

- Emphasis on biodiversity and optimal cultural practices.

The Disease Triangle

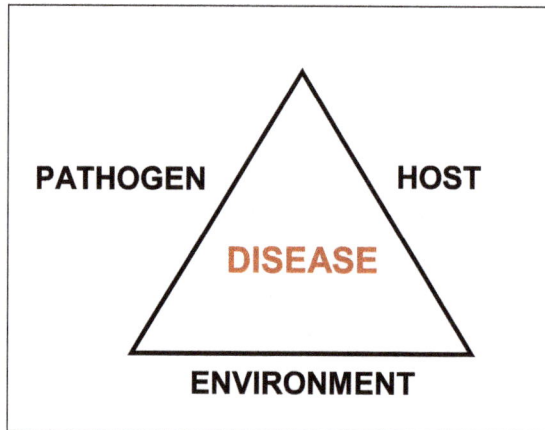

The disease triangle is a schematic that explains that in order to have disease you must have:

- The pathogen present on your farm. You can eliminate some pathogens through crop rotation, only using disease-free seed or plant stock, deep tillage to move soil pathogens deeper into the soil, etc.

- An appropriate host for the pathogen to live and reproduce. As you see, crop rotation would really help here. One of the most important crop protection methods is using disease resistant varieties. Make sure you are matching the correct disease resistance to the problem on your farm.

- Suitable environmental conditions are needed for disease to develop. Pathogens require environmental cues, such as humidity, moisture and desirable temperatures to germinate, survive and infect. If these conditions are not met, pathogens cannot survive. Ex. downy mildews require a certain amount of free moisture present on the leaf surface to germinate and infect. If you can increase air flow and the drying of the leaf surface you can make life miserable for the pathogen.

Disease Causing Organisms

- Fungi,

- Bacteria,

- Viruses,

- Nematodes.

Symptoms

- Fungi cause spots, lesions, blights, yellowing of leaves, wilts, cankers, rots, fruiting bodies, mildews, molds, leaf spots, root rots, cankers, and blotches. Fungi are typically spread by wind, rain, soil, mechanical means and infected plant material.

- Bacteria cause water-soaking, spots, wilts, rots, blights, cankers, exudates, galls, yellowing, leaf spots, watery blotches, wilting. Bacteria are typically spread by rain, mechanical means, planting material, vectors (ex. bacterial wilt of cucurbits spread by cucumber beetle).

- Viruses cause mottling, leaf and stem distortions, mosaic patterns, rings and stunting. Viruses cause interesting symptoms, some are beautiful. Viruses are spread by mechanical means, vectors and in plant material.

- Nematodes cause wilting, stunting, yellowing of entire plants. This is because the roots of the plant are infected and the plant is starving or thirsty. Nematodes are spread by soil on equipment or workers boots or on infected plant material.

Control

The NOP has a hierarchical approach to pest management starting with System-based cultural practices then mechanical and physical practices and finally material-based (chemical, botanical, elemental) practices.

Cultural Control

Cultural control is your first line of defense.

- Promote healthy soils and healthy plants. Healthy soil is the hallmark of organic agriculture. An unhealthy plant is very attractive to diseases.

- Soils rich in organic matter are shown to increase soil biodiversity and help to create and abundance of beneficial soil microorganisms. Using compost has been shown to increase the suppressiveness of the soil by encouraging beneficial microorganisms, as well as inducing disease resistance in plants by simply having healthier plants.

- Exclusion:
    ◦ Disease-free seeds, transplants or plant stock.
    ◦ Prevent introduction of diseased plants or soil.
    ◦ Disease free water source.
    ◦ Control insects that can carry disease.
    ◦ Soil solarization.
- Disease resistant varieties:
- Good sanitation from the prior season. Remove diseased plants or weeds from the field. Don't put disease plants or weeds in the compost pile. Many diseases are resistant to high heats or can become resistant. Some pathogens form resistant structures that can tolerate unfavorable conditions.
- Always work infested fields last and clean off equipment.
- Disinfest tools.
- Plant on raised beds. Not only helps with avoidance of pathogen, but also good moisture drainage is key.
- Crop rotation:
    ◦ >3 yrs between crops in the same family.
    ◦ Some pathogens cause disease among multiple plant families.
- Plants adapted to area.
- Plant at proper depth (below crown or graft).
- Use only thoroughly composted material.
- Improve air circulation by staking, trellis or pruning.
- Water in the morning.
- Avoid overhead irrigation if possible.

Make Life Difficult for the Pathogen

Create an unfavorable environment for pathogens:

- Increase air movement:
    ◦ Trellising, high tunnels.

- Increase soil drainage.

- Avoid low-lying areas.

- Row orientation:

    ○ Maximize air movement.

    ○ Minimize leaf wetness periods.

- Irrigation management:

    ○ Drip Irrigation.

- Mulches:

    ○ Plastic or plant-based.

- Reduce splash dispersal of pathogens.

- Protect fruit from soilborne pathogens.

- Avoidance:

    ○ Plant your crop when disease isn't as big a problem.

- Early blight and Cucurbit Downy Mildew.

Maximizing disease suppression with compost:

- Compost:

    ○ Cure 4 or more months.

    ○ Incorporate into soil several months before planting.

    ○ Inoculate with beneficial microorganisms, e.g. Trichoderma.

- Application:

    ○ 5-10 tons (dry weight)/A - rule of thumb.

    ○ Apply every year until significant organic matter improvement observed; watch for increases in P.

Produce Healthy Transplants

- Practice good sanitation in the greenhouse:

    ○ Use new or sanitized plug trays or flats and pathogen-free mixes.

    ○ Sanitize equipment – Install solid flooring; raise seedling trays.

- ○   Limit movement of personnel and equipment between greenhouses.

- ○   Clean benches, greenhouse structure thoroughly after the crop; close up greenhouse.

Variety Selection

Disease tolerance is the ability of a plant to endure an infectious or noninfectious disease, adverse conditions or chemical injury without serious damage or yield loss.

Disease resistance is when a plant possessed properties that prevent or impede disease development.

Pick varieties that are appropriate for your area. Keep records of a cultivar's performance and the disease pressure each season. Local heirlooms are generally better suited for a particular region. Use tissue culture plants (small fruits, some cut flowers, perennials) if available. These plants are often disease indexed. This is especially important for viruses. This is a specialized area and there are not always tissue culture plants available.

Physical and Mechanical Controls

Some options for physical and mechanical management of plant disease include:

- Hand-picking,
- Pruning,
- Mulches,
- Soil solarisation.

Pruning

- Prune out diseased plant parts,
- Increase light into canopy,
- Increase airflow,
- Helps spray penetrate all surfaces,
- Proper pruning for proper plant health.

Soil Solarisation

Used in greenhouses, seed beds, cold frames.

In greenhouses or in raised beds you can sterilize the soil or bench using heat produced by steam. You want to heat the coldest part of the soil to 82 °C for 30 minutes.

Soil solarization, a nonchemical technique, will control many soilborne pathogens and pests. This simple technique captures radiant heat energy from the sun, thereby causing physical, chemical, and biological changes in the soil. Transparent polyethylene plastic placed on moist soil during the hot summer months increases soil temperatures to levels lethal to many soilborne plant pathogens, weed seeds, and seedlings (including parasitic seed plants), nematodes, and some soil residing mites. Soil solarization also improves plant nutrition by increasing the availability of nitrogen and other essential nutrients.

Hot Water Seed Treatment

Research has shown that hot water seed treatment can help to decrease disease in seeds. Times and temps of seed treatment:

- Brussels sprouts, eggplant, spinach, cabbage, tomato = 122 °F for 25 min.

- Broccoli, cauliflower, cucumber, carrot, collard, kale, kohlrabi, rutabaga, turnip = 122 °F for 20 min.

- Mustard, cress, radish = 122 °F for 15 min.

- Pepper = 125 °F for 30 min.

- Lettuce, celery, celeriac – 118 °F for 30 minutes.

It is important to note that cucurbit seeds can be damaged by hot water. Other cautions include:

- Use new, high quality seed.

- Treat a small sample first and test for germination.

- Treat close to time of planting (within weeks).

- Treat only once.

Material Control

Materials include:

- Elemental fungicides:
  - Copper and sulfur.
- Biofungicides/Microorganisms:
  - Ex. PlantShield, MycoStop, Companion.
- Particle Film Barriers – Ex. kaolin clay.

- Peroxides and Bicarbonates.

- Compost Teas.

Sulfur

- Used effectively for powdery mildew on most crops.

- Labeled for rusts (grape and bean), botrytis (onions), black spot (rose).

- pH adjustment.

- Component of Bordeaux mixture.

- Lime sulfur - protectant dust or spray to control some fungal or bacterial diseases.

- Helps control rust, powdery mildew (PM), brown rot.

Copper

- Controls some fungi and bacteria.

- Free Cu - Copper sulfate: Bordeaux mixture.

- Fixed Cu - copper hydroxide, copper oxide, copper oxychloride, copper octanoate.

- Kocides – restricted use, requires license, OMRI approved.

- Safer Garden fungicide – Cu 12% or 0.4% : rust, scab, brown spot, black spot, others:

    ◦ Nasty stuff, some certifiers won't let you use it.

Botanical or Horticultural Oils

Used successfully to control insects that spread disease. Especially viral diseases. Some are effective for fungi like powdery mildews and rust.

Biocarbonates and Peroxides

- Bicarbonates - Potassium Bicarbonate (baking soda):

    ◦ disrupts cell membrane K balance.

    ◦ PM Black spot, leaf spots, rusts for seed, transplants or established plants.

    ◦ Ex. Kaligreen.

- Peroxides:
  - Disinfest plant surface.
  - Pre-plant, plant dip, foliar spray.
  - Use on tools, trays, pots, surfaces.
  - Ex. Oxi-Date.

Antibiotics Antibiotics -Streptomycin sulfate – many brands for agricultural use to control bacteria, fireblight.

- Fertilome Fireblight spray : also for bacterial wilt, stem rot, leaf spots and crown gall.
- Tetracycline – fireblight.

Biofungicides or Microorganisms

- Antagonists/Competitors:
  - Trichoderma harzianum is the most researched.
- Antifungal properties – Bacillus spp.
- Plant growth aids – Healthy roots, soil exploration.
- Trichoderma:
  - Activate plant immune system.
- Bacillus pumilus.

Compost Teas

Compost tea, in modern terminology, is a compost extract brewed with a microbial food source—molasses, kelp, rock dust, humic-fulvic acids. The compost-tea brewing technique, an aerobic process, extracts and grows populations of beneficial microorganisms.

Pre-plant Options

- Biofumigation:
  - Mustards, broccoli residue.
  - Muscodor – Broad-spectrum activity.
- Biocontrols – Contans, Advan LLC.

- Coniothyrium minitans a fungi used pre-plant.

- Narrow-spectrum (Sclerotinia only).

- Ex. lettuce drop, sclerotinia blight on peanut.

Scouting for Disease

Identify disease problems during the season:

- To change practices this year.

- For next year.

- Scout your crops on a regular basis (calendar). Scouting supplies include:

  ○ Hand lens (10X or higher).

  ○ Paper for notes.

  ○ Self sealing bags for samples.

  ○ A marker or pen – field guide.

  ○ Digital camera.

Using the Plant Disease and Insect Clinic: General Guidelines

- Collect fresh, don't send over weekend.

- Send several examples.

- Crush proof container.

- Provide lots of information.

- Don't pull plants-do bag roots.

- Don't get soil on foliage.

- Press leaves.

- Contact lab.

- Use your extension agent for help.

- Collect as much information as possible.

Early Blight: Potato and Tomato

Early blight is caused by two fungi (Alternaria solani and Alternaria tomatophila) that are a serious problem in tomatoes and potatoes but rarely effects peppers and egg-plants. All of the above-ground portions of the plant can be affected throughout the growing season. The disease starts on the lower leaves with small circular spots that have a target appearance of concentric rings. Leaves develop yellow blighted areas and later the tomato fruit may rot on the stem end. Potato tubers can also become infected, but this is quite rare. The pathogen can overwinter in the soil on diseased plant residues.

Cultural Control

- Use crop rotations of at least 3 years to non-hosts (away from tomato, potato and eggplant).

- Provide optimum growing conditions and fertility. Stressed plants (including drought) are more susceptible to early blight.

- Stake or cage plants to keep fruit and foliage away from soil.

- Drip irrigation is preferred, or overhead irrigation starting before dawn, so that the plants are dry early in the day. The key is to keep the period of leaf wetness to a minimum.

- Mulching helps to prevent splashing of spores from soil up to lower leaves.

- Indeterminate tomato and late-maturing potato varieties are usually more re-sistant/tolerant to early blight.

- Early blight can be seed-borne, so buy from a reliable supplier. Hot water seed treatment at 122 °F for 25 minutes is recommended to control early blight on tomato seed.

- Disinfect stakes or cages with an approved product each season before using. Sodium hypochlorite at 0.5% (12x dilution of household bleach) is effective, and must be followed by rinsing, and proper disposal of solution. Hydrogen perox-ide is also permitted.

Materials Approved for Organic Production

- Copper products showed one good and one poor result in recent studies.

- A Trichoderma harzanium product, PlantShield HC, used as a drench at planting, showed fair to good results in NY state on tomatoes over three seasons.

- Serenade, Bacillus subtilis. A protectant. Labeled, but considered only partially effective in UNH trails.

Downy Mildews

- Not true fungi:
    - ○ Watermolds,
    - ○ Swimming spores.
- Like cool wet weather.
- Overwinter as resistant spores in soil or infected plant material or blow in seasonally from diseased plantings.
- Effects vegetables and perennials, especially important in cucurbits, grapes and hops – Important to note that downy mildews are very host specific. Ex. the downy mildew that infects cucurbits is not the same as the one that infects grapes.

Cultural Control

- Resistant varieties.
- Planting date:
    - ○ We will get it in fall.
- Avoid overhead irrigation.
- Forecasting site.
- If transplanting- make sure transplants are disease-free.
- Rotation important for overall plant health.
- High tunnels.

Materials for Control of Downy Mildews

- OMRI-listed products:
    - ○ Copper, neem oil, biofungicides (ex. Serenade or Sonata), peroxides (ex. OxiDate), and bicarbonates (ex. Kaligreen).
- Compost teas.
- Best option as an organic grower is to use a copper product:
    - ○ Spray early in the morning to avoid phytotoxicity.
    - ○ Spraying copper prior to disease development or at very early onset (very few, mild symptoms), may help suppress the disease, but will not offer 100% control under favorable conditions (cool, wet and humid weather).

Powdery Mildew

Effects many plants, but like downy mildew powdery mildews are host specific. Powdery mildews like it hot and relatively dry (humid but not wet). Powdery mildew is perhaps one of the easiest diseases to diagnose.

Cultural Control

Resistant varieties:

- Plant in sunny areas with good air circulation.

- Avoid overhead irrigation.

- Avoid excess fertilization:

  ◦ Slow release better.

- Remove infected plant material.

Materials for Powdery Mildews

- Sulfur is very effective.

- Kaligreen and Armicarb (potassium bicarbonate-baking soda); dilute solutions of hydrogen peroxide (Oxidate):

  ◦ These materials burn out the fungus growing on the surface, but do not provide protection against new infections; thus, repeated applications are important.

- Oils:

  ◦ Saf-T-Side Spray Oil, Sunspray Ultra-Fine Spray Oil, or one of the plant-based oils such as neem oil or jojoba oil (e.g., E-rase).

  ◦ Be careful some plants are sensitive, esp. when used in conjunction with sulfur.

- Dilutions of milk and whey (the dairy by-product) have been effective for controlling powdery mildew (Australia).

Integrated Pest Management

Integrated pest management (IPM), also known as integrated pest control (IPC) is a broad-based approach that integrates practices for economic control of pests. IPM aims to suppress pest populations below the economic injury level (EIL). The UN's Food and Agriculture Organization defines IPM as "the careful consideration of all available pest control techniques and subsequent integration of appropriate measures that discourage the development of pest populations and keep pesticides and other interventions to levels that are economically justified and reduce or minimize risks to human health and the environment. IPM emphasizes the growth of a healthy crop with the least possible disruption to agro-ecosystems and encourages natural pest control mechanisms." Entomologists and ecologists have urged the adoption of IPM pest control since the 1970s. IPM allows for safer pest control.

An IPM boll weevil trap in a cotton field.

The introduction and spread of invasive species can also be managed with IPM by reducing risks while maximizing benefits and reducing costs.

Applications

IPM is used in agriculture, horticulture, forestry, human habitations, preventive conservation and general pest control, including structural pest management, turf pest management and ornamental pest management.

Principles

An American IPM system is designed around six basic components:

- Acceptable pest levels — The emphasis is on *control*, not *eradication*. IPM holds that wiping out an entire pest population is often impossible, and the attempt can be expensive and unsafe. IPM programmes first work to establish acceptable pest levels, called action thresholds, and apply controls if those thresholds are crossed. These thresholds are pest and site specific, meaning that it may be acceptable at one site to have a weed such as white clover, but not at another site. Allowing a pest population to survive at a reasonable threshold reduces selection pressure. This lowers the rate at which a pest develops resistance to a control, because if almost all pests are killed then those that have resistance will provide the genetic basis of the future population. Retaining a significant number of unresistant specimens dilutes the prevalence of any resistant genes that appear. Similarly, the repeated use of a single class of controls will create pest populations that are more resistant to that class, whereas alternating among classes helps prevent this.

- Preventive cultural practices — Selecting varieties best for local growing conditions and maintaining healthy crops is the first line of defense. Plant quarantine and 'cultural techniques' such as crop sanitation are next, e.g., removal of diseased plants, and cleaning pruning shears to prevent spread of infections. Beneficial fungi and bacteria are added to the potting media of horticultural crops vulnerable to root diseases, greatly reducing the need for fungicides.

- Monitoring — Regular observation is critically important. Observation is broken into inspection and identification. Visual inspection, insect and spore traps, and other methods are used to monitor pest levels. Record-keeping is essential, as is a thorough knowledge of target pest behavior and reproductive cycles. Since insects are cold-blooded, their physical development is dependent on area temperatures. Many insects have had their development cycles modeled in terms of degree-days. The degree days of an environment determines the optimal time for a specific insect outbreak. Plant pathogens follow similar patterns of response to weather and season.

- Mechanical controls — Should a pest reach an unacceptable level, mechanical methods are the first options. They include simple hand-picking, barriers, traps, vacuuming and tillage to disrupt breeding.

- Biological controls — Natural biological processes and materials can provide control, with acceptable environmental impact, and often at lower cost. The main approach is to promote beneficial insects that eat or parasitize target pests. Biological insecticides, derived from naturally occurring microorganisms(e.g.—Bt, entomopathogenic fungi and entomopathogenic nematodes), also fall in this category. Further 'biology-based' or 'ecological' techniques are under evaluation.

- Responsible use — Synthetic pesticides are used as required and often only at specific times in a pest's life cycle. Many newer pesticides are derived from plants or naturally occurring substances (e.g.—nicotine, pyrethrum and insect juvenile hormone analogues), but the toxophore or active component may be altered to provide increased biological activity or stability. Applications of pesticides must reach their intended targets. Matching the application technique to the crop, the pest, and the pesticide is critical. The use of low-volume spray equipment reduces overall pesticide use and labor cost.

An IPM regime can be simple or sophisticated. Historically, the main focus of IPM programmes was on agricultural insect pests. Although originally developed for agricultural pest management, IPM programmes are now developed to encompass diseases, weeds and other pests that interfere with management objectives for sites such as residential and commercial structures, lawn and turf areas, and home and community gardens.

Process

IPM is the selection and use of pest control actions that will ensure favourable economic condition, ecological and social consequences and is applicable to most agricultural, public health and amenity pest management situations. The IPM process starts with monitoring, which includes inspection and identification, followed by the establishment of economic injury levels. The economic injury levels set the economic threshold level. That is the point when pest damage (and the benefits of treating the pest) exceed the cost of treatment. This can also be an action threshold level for determining an unacceptable level that is not tied to economic injury. Action thresholds are more common in structural pest management and economic injury levels in classic agricultural pest management. An example of an action threshold is one fly in a hospital operating room is not acceptable, but one fly in a pet kennel would be acceptable. Once a threshold has been crossed by the pest population action steps need to be taken to reduce and control the pest. Integrated pest management employs a variety of actions including cultural controls such as physical barriers, biological controls such as adding and conserving natural predators and enemies of the pest, and finally chemical controls or pesticides. Reliance on knowledge, experience, observation and integration of multiple techniques makes IPM appropriate for organic farming (excluding synthetic pesticides). These may or may not include materials listed on the Organic Materials Review Institute (OMRI) Although the pesticides and particularly insecticides used in organic farming and organic gardening are generally safer than synthetic pesticides, they are not always more safe or environmentally friendly than synthetic pesticides and can cause harm. For conventional farms IPM can reduce human and environmental exposure to hazardous chemicals, and potentially lower overall costs.

Risk assessment usually includes four issues: 1) characterization of biological control agents, 2) health risks, 3) environmental risks and 4) efficacy.

Mistaken identification of a pest may result in ineffective actions. E.g., plant damage due to over-watering could be mistaken for fungal infection, since many fungal and viral infections arise under moist conditions.

Monitoring begins immediately, before the pest's activity becomes significant. Monitoring of agricultural pests includes tracking soil/planting media fertility and water quality. Overall plant health and resistance to pests is greatly influenced by pH, alkalinity, of dissolved mineral and oxygen reduction potential. Many diseases are water-borne, spread directly by irrigation water and indirectly by splashing.

Once the pest is known, knowledge of its lifecycle provides the optimal intervention points. For example, weeds reproducing from last year's seed can be prevented with mulches and pre-emergent herbicide.

Pest-tolerant crops such as soybeans may not warrant interventions unless the pests are numerous or rapidly increasing. Intervention is warranted if the expected cost of damage by the pest is more than the cost of control. Health hazards may require intervention that is not warranted by economic considerations.

Specific sites may also have varying requirements. E.g., white clover may be acceptable on the sides of a tee box on a golf course, but unacceptable in the fairway where it could confuse the field of play.

Possible interventions include mechanical/physical, cultural, biological and chemical. Mechanical/physical controls include picking pests off plants, or using netting or other material to exclude pests such as birdsfrom grapes or rodents from structures. Cultural controls include keeping an area free of conducive conditions by removing waste or diseased plants, flooding, sanding, and the use of disease-resistant crop varieties. Biological controls are numerous. They include: conservation of natural predators or augmentation of natural predators, sterile insect technique (SIT).

Augmentation, inoculative release and inundative release are different methods of biological control that affect the target pest in different ways. Augmentative control includes the periodic introduction of predators. With inundative release, predators are collected, mass-reared and periodically released in large numbers into the pest area. This is used for an immediate reduction in host populations, generally for annual crops, but is not suitable for long run use. With inoculative release a limited number of beneficial organisms are introduced at the start of the growing season. This strategy offers long term control as the organism's progeny affect pest populations throughout the season and is common in orchards. With seasonal inoculative release the beneficials are collected, mass-reared and released seasonally to maintain the beneficial population. This is commonly used in greenhouses. In America and other western countries, inundative releases are predominant, while Asia and the eastern Europe more commonly use inoculation and occasional introductions.

The sterile insect technique (SIT) is an area-wide IPM program that introduces sterile male pests into the pest population to trick females into (unsuccessful) breeding encounters, providing a form of birth controland reducing reproduction rates. The biological controls mentioned above only appropriate in extreme cases, because in the introduction of new species, or supplementation of naturally occurring species can have detrimental ecosystem effects. Biological controls can be used to stop invasive species or pests, but they can become an introduction path for new pests.

Chemical controls include horticultural oils or the application of insecticides and herbicides. A green pest management IPM program uses pesticides derived from plants, such as botanicals, or other naturally occurring materials.

Pesticides can be classified by their modes of action. Rotating among materials with different modes of action minimizes pest resistance.

Evaluation is the process of assessing whether the intervention was effective, whether it produced unacceptable side effects, whether to continue, revise or abandon the program.

Southeast Asia

The Green Revolution of the 1960s and '70s introduced sturdier plants that could support the heavier grain loads resulting from intensive fertilizer use. Pesticide imports by 11 Southeast Asian countries grew nearly sevenfold in value between 1990 and 2010, according to FAO statistics, with disastrous results. Rice farmers become accustomed to spraying soon after planting, triggered by signs of the leaf folder moth, which appears early in the growing season. It causes only superficial damage and doesn't reduce yields. In 1986, Indonesia banned 57 pesticides and completely stopped subsidizing their use. Progress was reversed in the 2000s, when growing production capacity, particularly in China, reduced prices. Rice production in Asia more than doubled. But it left farmers believing more is better—whether it's seed, fertilizer, or pesticides.

The brown planthopper, *Nilaparvata lugens*, the farmers' main target, has become increasingly resistant. Since 2008, outbreaks have devastated rice harvests throughout Asia, but not in the Mekong Delta. Reduced spraying allowed natural predators to neutralize planthoppers in Vietnam. In 2010 and 2011, massive planthopper outbreaks hit 400,000 hectares of Thai rice fields, causing losses of about $64 million. The Thai government is now pushing the "no spray in the first 40 days" approach.

By contrast early spraying kills frogs, spiders, wasps and dragonflies that prey on the later-arriving and dangerous planthopper and produced resistant strains. Planthoppers now require pesticide doses 500 times greater than originally. Overuse indiscriminately kills beneficial insects and decimates bird and amphibian populations. Pesticides are suspected of harming human health and became a common means for rural Asians to commit suicide.

In 2001, scientists challenged 950 Vietnamese farmers to try IPM. In one plot, each farmer grew rice using their usual amounts of seed and fertilizer, applying pesticide as they chose. In a nearby plot, less seed and fertilizer were used and no pesticides were applied for 40 days after planting. Yields from the experimental plots was as good or better and costs were lower, generating 8% to 10% more net income. The experiment led to the "three reductions, three gains" campaign, claiming that cutting the use of seed, fertilizer and pesticide would boost yield, quality and income. Posters, leaflets, TV commercials and a 2004 radio soap opera that featured a rice farmer who gradually accepted the changes. It didn't hurt that a 2006 planthopper outbreak hit farmers using insecticides harder than those who didn't. Mekong Delta farmers cut insecticide spraying from five times per crop cycle to zero to one.

The Plant Protection Center and the International Rice Research Institute (IRRI) have been encouraging farmers to grow flowers, okra and beans on rice paddy banks, instead of stripping vegetation, as was typical. The plants attract bees and a tiny wasp that eats planthopper eggs, while the vegetables diversify farm incomes.

Agriculture companies offer bundles of pesticides with seeds and fertilizer, with incentives for volume purchases. A proposed law in Vietnam requires licensing pesticide dealers and government approval of advertisements to prevent exaggerated claims. Insecticides that target other pests, such as Scirpophaga incertulas (stem borer), the larvae of moth species that feed on rice plants allegedly yield gains of 21% with proper use.

Disease Management in Organic Lettuce Production

Lettuce (Lactuca sativa L.) is the world's most popular leafy salad vegetable. Various types of lettuce are cultivated across the world, primarily for human consumption of their fresh leaves. Among these types of lettuce are leaf (loose-leaf lettuce), Cos (romaine), crisphead (iceberg), butterhead, stem (asparagus lettuce), and numerous others, which have limited production. Lettuce responds well to a moist, rich soil, full exposure to sun and cool weather conditions. It is typically a crop of temperate climates. It is grown everywhere where the average temperature remains between 45 °F and 65 °F during the growing season. Although primarily adapted to colder climates, there are summer and winter varieties that can be used in other seasons. Sandy peat and muck, deep black sandy loams and loams are the most suitable types of soil. Lettuce is slightly tolerant to acidic soils, but the ideal pH ranges between 6.0 and 6.8.

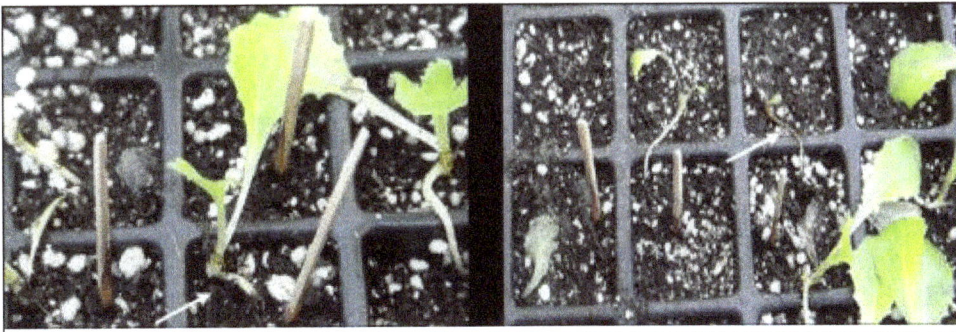

Damping-off (Rhizoctonia solani).

One of the major challenges facing organic lettuce producers is disease management. The losses in lettuce production due to disease can be significant and devastating under favorable conditions. The primary lettuce diseases are bottom rot (Rhizoctonia solani), damping-off (Pythium spp. and R. solani), downy mildew (Bremia lactucae), drop (Sclerotinia sclerotiorum and S. minor), gray mold (Botrytis cinerea), powdery mildew (Erysiphe cichoracearum), Septoria leaf spot (Septoria lactucae), bacterial leaf spot (Xanthomonas campestris pv. vitians), soft rot (Pectobacterium carotovorum subsp. carotovorum), aster yellows phytoplasma, lettuce mosaic virus (LMV), Turnip mosaic virus (TuMV), Beet western yellows virus (BWYV) and Tomato spotted wilt virus (TSWV). Preventive measures against such diseases have priority in organic lettuce production.

Downy mildew (Bremia lactucae).

Gray mold (Botrytis cinerea).

Powdery mildew (Erysiphe cichoracearum).

Bacterial leaf spot (Xanthomonas campestris pv. vitians).

Disease Management Strategies

Seed Selection and Seed Health Treatments

The planting of pathogen-free organically produced lettuce seed is an important first step in managing diseases. There are some seed treatments, such as hot-water sanitation or National Organic Program (NOP)-compliant protectants that can be used by organic farmers to eradicate some pathogens from seed. Hot-water seed treatments (118 °F for 30 min) can greatly decrease seedborne inoculum of some disease such as Septoria leaf spot and bacterial leaf spot. Treatments of lettuce seed with solutions of aqueous 3 to 5% hydrogen peroxide may also effectively reduce or eradicate the bacterial leaf spot pathogen (X. campestris pv. vitians) from heavily infested lettuce seed. However, hydrogen peroxide treatments at a concentration of 5% reduced seed germination up to 28% compared with controls. On the other hand, small but significant reductions in germination were observed on seed lots treated with 3% hydrogen peroxide. Treatment of lettuce seed with solutions of 0.52% sodium hypochlorite for 5 min soaking time also reduced bacterial leaf spot. However, chlorine seed treatment of organic seed must be followed by a rinse with potable water. Check with your suppliers of organic seed regarding any seed treatments they apply or use to suppress seed-borne diseases. Also, growers should check with their certifying agency regarding any seed treatments that they intend to apply prior to planting.

Variety Selection

Many cultivars (varieties) of lettuce have been developed that are resistant and/or tolerant to specific diseases. The terms resistance or tolerance do not mean that the plant is completely immune to disease. They refer to a plant's ability to overcome, to some degree, the effect of the pathogen. Also, no variety is resistant or tolerant to all diseases. Resistant varieties should always be used in combination with other management practices.

Organic Amendments

The addition of organic matter such as good quality compost can aid in reducing diseases caused by soilborne pathogens. Organic matter improves soil structure and its ability to hold water and nutrients; it also supports microorganisms that contribute to biological control. The incidence of lettuce drop and survival rate of sclerotia of S. sclerotiorum can be reduced by the addition of stable manure (a mixture of straw, horse dung, and urine), fowl manure (a mixture of wood shavings and chicken droppings), and Lucerne hay to the field. Note: In order to be used for organic production, animal manure must be fully composted or else incorporated into the soil at least 120 days prior to the harvest of organic lettuce. Incorporating approved amendments, such as blood meal, fish meal, or feather meal, into soil can also increase the marketable yield and quality of lettuce.

Propagation and Nursery Management

Lettuce can be either transplanted or direct seeded, depending mostly on cost and availability of transplants and the time of year. Use of transplants will help manage some diseases. Direct seeding depths are about 1/4 to 3/8 inch. In case of heavier soils, shallower depth is recommended to reduce the risk of damping-off disease problems. Raised beds are ideal for lettuce production. They help prevent damage from soil compaction and flooding. They also improve airflow around the plants resulting in reduced disease incidence. Over-watering or planting in poorly drained soils must be avoided to prevent root diseases and seed decay. Plants should be irrigated without applying water to the foliage, which helps reduce most foliar lettuce diseases, particularly bacterial leaf spot. In research on irrigation methods, lettuce drop incidence was significantly lower and yields were higher in plots under subsurface drip irrigation compared with furrow irrigation. In another study, subsurface drip irrigation reduced human health risks when microbial-contaminated water was used for irrigation.

Sanitation

Many pathogens survive between crops in or on the residue from diseased plants, so it is important to remove as much of the old plant debris as possible. When lettuce drop is observed in the production area, the diseased plants and soil immediately surrounding the infected plant should be removed. Weeds should be eliminated as they may harbor pathogens or serve as hosts for insects that transmit viruses, aster yellows phytoplasma and other pathogens. Frequent disinfection of harvest tools will also help prevent the spread of soft rot, post harvest pathogens such as gray mold, and microbial contaminants that may be human pathogens.

Rotation

Crop rotation has been used successfully for disease management in different pathosystems. Hence, it has drawn increased attention as an important disease management tool in organic agriculture. A good rotation scheme is to treat members of the same plant family as a group and rotate based on groups rather than individual crops. For example, lettuce should be rotated with vine crops, tomatoes, or cole crops but not other types of lettuce, chicory or endive. It is important to note that one difficulty in using crop rotation to control some diseases is the limited availability of crops for rotation. For example, Sclerotinia on lettuce cannot be managed with rotation because most crops are hosts of Sclerotinia and the pathogen can survive for long periods between host crops. Broccoli (Brassica oleracea), used as a green manure, recently has been shown to reduce the number of S. minor sclerotia in soil.

Row Covers

Floating row covers are an excellent barrier to some early pests that vector disease, such as leafhoppers, thrips, whiteflies and aphids. Row covers are made of lightweight fabric that can be laid directly over plants or supported with hoops. The fabric needs to be secured to keep pests out. This can be accomplished by using metal staples made for this purpose. Row covers should be removed when temperatures regularly reach the high 80's.

Materials

Early detection is important to manage diseases. It is therefore important to scout plants regularly and know which diseases are present in the crop. When preventive and cultural methods for disease control are insufficient to manage a disease, National Organic Program (NOP) compliant inputs can be applied.

Before applying any pest control product, be sure to 1) read the label to be sure that the product is labeled for the crop and the disease you intend to control, 2) read and understand the safety precautions and application restrictions, and 3) make sure that the brand name product is listed in your Organic System Plan and approved by your certifier.

Copper oxide can be used against lettuce downy mildew disease. However, over-application can lead to copper accumulation in the soil, contamination of run-off water, and toxicity to non-target organisms. Sulfur products can provide control for powdery mildew. Biorational products represent an important option for the management of plant diseases. On lettuce, Contans (Coniothyrium minitans) has been effective in reducing the incidence of lettuce drop by as much as 50% in greenhouse crops. In another study, two applications of Contans, one at planting and one at post-thinning, controlled lettuce drop caused by S. sclerotiorum in a desert lettuce production system.

Disease Management in Organic Seed Production

Diseases can have a significant effect on production of specialty seed crops. Seed growers must pay attention to diseases that affect the vegetative growth stage of the crop, as well as those that affect the reproductive growth stages (flowering and seed formation). Some diseases, such as Verticillium wilt of spinach, become symptomatic only when the crop enters the reproductive stage; these diseases are more important to seed growers than to vegetable growers (unless the vegetable crop also has a flowering stage, e.g., tomato or potato). While vegetable growers are concerned primarily with the pathogens that affect marketable yield and quality, seed growers must also learn how to diagnose and manage seedborne pathogens and the microorganisms that affect seed quality. Pathogens usually remain viable for longer in seed than in vegetative parts of the plant or in the soil. Seeds are a major means of survival of some plant pathogens and of introducing new pathogens to a field or region.

Disease management tactics are either preventive (actions taken to avoid or reduce the likelihood of disease problems) or curative (treatments that eliminate or reduce the effects of a particular disease after it has become established). Because there are few effective curative practices available to organic farmers, organic farmers focus their disease management efforts primarily on preventive cultural practices. Such practices include planting pathogen-free seed, planting in fields of low inoculum potential and in locations with good air movement, adopting wide row spacing, orienting the crop rows to maximize air movement between rows, and tying or staking seed crops to improve air circulation and reduce humidity in the canopy. If feasible, consider using drip or furrow irrigation instead of overhead irrigation, or irrigate earlier in the day to allow the canopy to dry before nightfall.

Some significant pathogens of seed crops are soilborne, such as Fusarium wilt of spinach. To manage soilborne pathogens, it is important to know the cropping history of the field and to adopt appropriate crop rotations. A rotation of 6 to 15 years, depending on the susceptibility of the spinach cultivar, is required to control Fusarium wilt in spinach seed crops. Some soilborne pathogens affect more than one crop, e.g., the fungus that causes Verticillium wilt of spinach can also infect potato, so it is important to avoid growing other crops in the rotation that may be alternative hosts to soilborne pathogens that affect the seed crop.

Strict management of, and screening for, seedborne pathogens of vegetable crops is critical to maintaining high seed quality. Even low levels of seed contamination can cause epidemics of some diseases when infected seed is planted in the field. For example, the tolerance level for contamination of crucifer seed with the causal agent of black rot, Xanthomonas campestris pv. campestris, is 0 contaminated seeds in 10,000 to 50,000 seeds (depending on the market or country in which the seed will be distributed).

Seeds contaminated with a pathogen can be treated physically (e.g. hot water) or chemically (e.g. bleach) to destroy inoculum or reduce the incidence of infection. Some physical and chemical treatments may reduce seed quality (germination, vigor, and/or longevity), so it may be important to test a particular seed treatment on a small sample of seed and check for possible phytotoxicity to the seed before treating an entire seed lot. Hot water treatment can only be used on some crops, such as brassicas, carrots, tomatoes, peppers, and lettuce, but even on those crops very precise parameters must be followed for hot water treatment to avoid damaging the seed. There are a number of biological and natural disease management products coming onto the market that are approved for use on organic farms, but it must be noted that the efficacy of these biocontrol products may vary among sites, crops, and diseases, reflecting the the complexities and particulars of interactions amongst the host, pathogen and environment. Therefore, planting pathogen-free seed, when possible, is always preferable to trying to eradicate a pathogen from seed.

Keys to Disease Management in Organic Seed Crops

Three ingredients are necessary for a disease to develop in your seed crop:

- The pathogen that causes the disease must be present.

- The host plant must be susceptible (not resistant).

- The environment (weather, microclimate) must be conducive (must support infection of the plant, growth of the pathogen).

Consider all of these ingredients when developing a disease management plan.

Pathogen

- Know the pathogen: Learn about, scout for and diagnose diseases on your crops.

- Exclude the pathogen: don't let it arrive on your farm. Don't bring in diseased transplants or seed. Don't grow your crop in the vicinity of other diseased fields to prevent wind or rain dispersal of the pathogen. Don't bring the pathogen in from other fields on people or equipment. Know which seedborne pathogens are important to your crop and prevalent in your region, and ensure, through communication with your seed supplier or contractor, that the seed you are planting has been tested to be pathogen-free or has been preventatively treated.

- Avoid the pathogen: Don't plant your crop in a field that contains the pathogen, for example, a soil that contains inoculum of Verticillium or Sclerotinia. Practice appropriate crop rotation to reduce the amount of inoculum in the field before again planting a susceptible crop. Some soilborne pathogens affect more than one crop, e.g., the fungus that causes Verticillium wilt of spinach can also infect potato, so it is important to avoid growing other crops in the rotation that may be alternative hosts to soilborne pathogens. While most soilborne diseases can be managed with a 3-5 year rotation, this is not true for some very important seed crop diseases. For example, a rotation of 6 to 15 years, depending on the susceptibility of the spinach cultivar, is required to control Fusarium wilt in spinach seed crops.

- Eradicate the pathogen: Destroy diseased crop residues, cull piles, and alternate hosts to the pathogen. Scout fields regularly for early symptoms of disease development. If feasible, rogue the symptomatic plants, and remove them from the field.

- Treat seeds for the pathogen: Treat seed with hot water or other organic treatments to eliminate the pathogen. Seeds contaminated with a pathogen can be treated physically (e.g. hot water), or chemically (e.g. bleach) to destroy inoculum or reduce the incidence of infection. Some physical and chemical treatments may reduce seed quality (germination, vigor, and/or longevity), so it may be important to test a particular seed treatment on a small sample of seed and check for possible damage to the seed before treating an entire seed lot. Hot water treatment can be used only on some crops, such as brassicas, carrots, tomatoes, peppers, and lettuce. Even on those crops very precise parameters must be followed to avoid damaging the seed. When possible, planting pathogen-free seed is alway preferable to trying to eradicate a pathogen from seed.

Host Plant

- Therapy: Strict management of, and screening for, seedborne pathogens of vegetable crops is critical for maintainence of high seed quality. Low levels of contaminated seed can cause epidemics of some diseases when infected seed is planted. For example, the tolerance level for contamination of crucifer seed with the causal agent of black rot, Xanthomonas campestris pv. campestris, is one contaminated seed in 10,000 to 50,000 seeds, depending on the market or country in which the seed will be distributed. This level is as important when obtaining the seed lot to grow out as it is when selling a seed lot. Typically, it is the seed company's responsibility to screen seed lots for pathogens before sale. All diagnostic laboratory results should be communicated to the seed grower. Commercial seed health tests are not available for all seedborne pathogens, particularly specialty crops grown on a small scale, and for which there has been limited seedborne pathogen research.

- Host resistance: Select resistant cultivars. When available, this is the easiest and most reliable method to control seedborne plant diseases. Many seed catalogs and extension publications provide information on resistance and susceptibility of specific cultivars to common diseases.

Wild Garden Seed lettuce disease resistance trial. Frank Morton of Wild Garden Seed conducted a trial of many lettuce cultivars over several years to screen them for disease resistance and tolerance.

- Protection: Paint materials on a plant to reduce the success of the pathogen and slow down disease progress (for example, apply copper or sulfur). Organic farmers must describe in their organic system plan their rationale for applying a pest control material. Communicate with your certifier to ensure any materials applied to crops for disease control are permitted for use on organic farms. Always ensure that the product is labeled for the crop and disease of concern. There are a number of biological and natural disease management products on the market that are approved for organic production, but their efficacy is not well documented and it may vary by site. This is a reflection of the complexity of soil - climate systems.

Environment

- Avoid environmental conditions that promote infection and disease development.

- Use drip irrigation or only irrigate early in the morning. Plants wetted in the morning will dry quckly in the afternoon sun, reducing the length of time that leaves remain wet. During the summer, most pathogens require 8 to 12 hours of wetness to initiate an infection.

- Weed your field and stake your plants to improve air flow through the canopy. Good airflow speeds drying of the canopy and reduces the length of wet leaf conditions.

Wild Garden Seed irrigates their lettuce seed fields with drip lines and keeps them weed-free to minimize leaf wetness and disease development.

- Plant your crop earlier or later in the season to avoid weather conditions that encourage disease development. Grow your crop in high tunnels to control environmental variables.

- Use geotextile fabric to keep plant materials dry as seeds mature.

Wild Garden Seed uses geotextile fabric to keep plants and seeds dry as they mature in the field. They also use geotextile fabric to safely and easily move the seed during processing.

Early Blight Management for Organic Tomato Production

Early blight is caused by Alternaria solani and A. tomatophila, which survive between crops on infected crop residues and on solanaceous host weeds. These fungi can also be carried on tomato seed. Early blight is common on tomatoes and potatoes, and it occasionally infects eggplants and peppers. It causes direct losses by the infection of fruits and indirect losses through leaf lesions, which reduce plant vigor. The disease is favored by warm temperatures and extended periods of leaf wetness from frequent rain, overhead irrigation, or dew. The fungal spores can be spread by wind and rain, irrigation, insects, workers, and on tools and equipment. Once the primary infections have occurred, they become the most important source of new spore production and are responsible for rapid disease spread. Early blight can develop quickly mid- to late season and is more severe when plants are stressed by poor nutrition, drought, other diseases, or pests.

Symptoms

Early blight on tomato leaf.

Early blight on tomato stems.

Early blight first appears as small brown-to-black lesions on older foliage. The tissue surrounding the primary lesions may become bright yellow, and when lesions are numerous, entire leaves may become chlorotic. As the lesions enlarge, they often develop concentric rings giving them a bull's-eye or target-spot appearance. When conditions are favorable for disease development, lesions can become numerous and plants defoliate, reducing both fruit quantity and quality. Fruit can become infected either in the green or ripe stage through the stem attachment. Lesions can become quite large, involve the whole fruit, and have characteristic concentric rings. Infected fruit often drops, and losses of immature fruit may occur. Fruit on defoliated plants are also subject to sunscald. Stems and petioles affected by early blight have elliptical concentric lesions, which severely weaken the plant. Lesions at the base of emerging seedlings can cause collar rot. If this arises consecutively on many seedlings, it may indicate contamination of tomato seeds or soil used for planting.

Management

Seed Selection and Treatments

The planting of pathogen-free organically produced tomato seed is an important first step in managing early blight disease. Fungicidal seed treatments are not an option for organic growers; however, there are some seed treatments—such as hot water sanitation or National Organic Program (NOP)-compliant protectants—that can be used by organic farmers to eradicate some pathogens from seed. Hot water seed treatment at 122°F for 25 minutes is recommended to control early blight on tomato seed. Chlorine seed treatment is not an acceptable treatment. Growers should always check with their certifying agency prior to using any seed treatment.

Variety Selection

A number of tomato varieties have been developed that are partially resistant to early blight, such as the Mountain series—including Mountain Pride, Mountain Supreme, Mountain Gold, Mountain Fresh, and Mountain Belle. Such varieties are sometimes inaccurately described as "tolerant" to early blight. Partially resistant varieties are not immune to a particular disease; in the case of partial resistance to early blight, infected plants may develop lesions or defoliate to some degree, often depending on environmental conditions, plant stress, and other factors.

Propagation and Nursery Management

The planting of pathogen-free organically produced tomato transplants is essential in managing early blight disease. Seeds should be sown in a pathogen-free planting mix that meets organic standards. Field soil may contain pathogens, weed seeds, and insect pests, so it is not recommended for seedling production. Flats with larger cells allow greater air movement between seedlings, which promote rapid drying of foliage and discourages disease development. Mulch (black plastic, straw, and newspaper, for example) helps to protect the plant from inoculum splashing from the soil onto lower leaves.

Crop Rotation

Early blight is a soilborne disease, so rotation can be a good management tool. A good practice is to treat members of the same plant family as a group and rotate based on groups rather than individual crops. Solanaceous crops include tomatoes, potatoes, peppers, chilies, eggplants, and tobacco. Using a three or four year crop rotation with non-solanaceous crops will allow infested plant debris to decompose in the soil. Rotations with small grains, corn, or legumes are preferable.

Organic Amendments

Good quality compost improves soil structure and its ability to hold water and nutrients; it also supports microorganisms that contribute to biological control. Our research has shown that early blight severity was less in tomato plants grown in compost-amended soil in the high tunnel than in non-amended soil; furthermore, incorporating the amendments into soil increased the total and marketable yield.

Sanitation

Early blight survives between crops in or on the residue from diseased plants, so it is important to remove diseased plants or destroy them immediately after harvest. Alternatively, bury diseased crop debris by deep-plowing to reduce spore levels available for infection of new plants. Solanaceous weeds, such as jimsonweed, horse nettle, ground-cherry, and the numerous nightshades, should be eliminated as they may harbor pathogen inoculum. Volunteer potatoes and tomatoes can also be a source of inoculum for early blight. Frequent disinfection of pruning tools should be disinfected frequently during use to help prevent the spread of spores. Stakes and cages can be disinfected each season before use with an approved product, such as ethanol, hydrogen peroxide, or peracetic acid. Disinfection with sodium hypochlorite (bleach) at 0.5% is effective, but must be followed by rinsing.

Materials

Early detection is important in disease management, so it is important to scout plants regularly. Disease forecasting is another important practice used to predict the probability of disease incidence. Weather monitoring instruments are placed in the field to collect data on canopy temperature, leaf wetness periods, and other factors that affect the likelihood of disease occurrence. The data collected from these monitoring stations are used to time fungicidal sprays for their optimum effect, generally resulting in fewer spray applications each growing season. If the uses of preventive and cultural methods for disease control are insufficient to manage early blight, National Organic Program (NOP)-compliant inputs can be applied to reduce disease spread. Before applying any pest control product, be sure to: (1) read the label to be sure that the product is labeled for the crop you intend to apply it to and the disease you intend to control, (2) read and understand the safety precautions and application restrictions, and (3) make sure that the brand name product is listed in your Organic System Plan and is approved by your certifier.

Copper products are considered synthetic and allowed with restrictions. Fixed copper products, hydrogen dioxide (hydrogen peroxide), and potassium bicarbonate can be used against early blight. However, over-application of copper products can lead to copper accumulation in the soil, contamination of run-off water, and subsequent toxicity to non-target organisms. Biorational products represent an important option for the management of plant diseases. Research has shown that A. solani-inoculated tomato plants treated with compost extract, prepared in a ratio of 1:5 compost:water (v/v), showed a significant reduction in disease index as compared with the untreated inoculated plants. Other research has shown that the efficacy of compost tea was improved when combined with the biofungicides Serenade Max (Bacillus subtilis) and Sonata (Bacillus pumilis). These results indicate that the use of compost tea for control of tomato early blight disease may be of some benefit to greenhouse tomato growers, and to growers of organic field tomatoes who are limited in their disease management options . Garlic- and neem oils and seaweed extract have also been shown to be effective in reducing the severity of early blight disease on tomato compared to untreated controls in another research project .

Management of Black Rot of Cabbage and other Crucifer Crops in Organic Farming Systems

The cabbage above shows typical black rot symptoms, with V-shaped lesions moving into the leaf from the leaf margin.

Black rot, caused by the bacterium Xanthomonas campestris pv. campestris (Xcc), is a significant disease of cabbage and other crucifer crops worldwide. The disease was first described in New York on turnips in 1893, and has been a common problem for growers for over 100 years. The pathogen thrives in warm, wet weather, spreading from plant to plant by splashing water, wind blown water droplets, and by workers or animals moving from infected fields to healthy fields. Xcc can spread rapidly during transplant production in greenhouses or seed beds, and could be spreading long before any symptoms are observed. The bacterium can infest seed, infecting young seedlings as they emerge. The pathogen can also survive in cruciferous weeds, such as yellow rocket, Shepherd's purse, and wild mustard, as well as in crop debris in the field.

Transplants with black rot symptoms are shown above. While these plants are clearly diseased, it is important to remember that bacteria can be invading plants even if no symptoms are observed.

Susceptible Crops

All crucifer crops are susceptible to black rot, including cabbage, broccoli, cauliflower, Brussels sprouts, Chinese cabbage, kale, radish, turnip, mustard, rutabaga, watercress, and arugula.

The cabbage field on the left has been destroyed by the black rot pathogen. Portions of the field on the right have been overtaken by cruciferous weeds which can serve as a source of inoculum.

Symptoms and Biology

Symptoms of black rot generally begin with yellowing at the leaf margin, which expands into the characteristic "V"-shaped lesion. The bacterium commonly enters the plant through the hydathode, or water pore, on the margin of the leaf; however, damage to leaves due to insect feeding, hail, or mechanical injury can also enable pathogen entry. The bacterial infection becomes systemic, meaning that the bacterium can enter the veins of the plant and spread into the cabbage head, which can lead to serious losses in storage. Blackening of the vascular tissue is typical in severe infections.

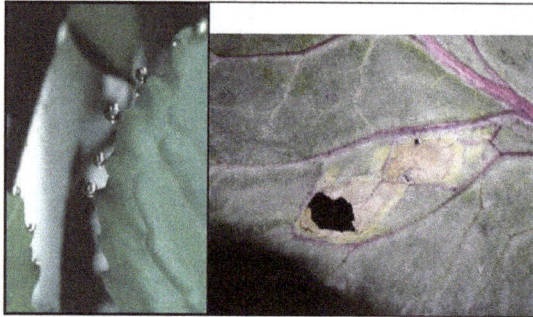

Hydathodes (or pores) on the margin of this cabbage leaf (left) exude plant sap or guttation droplets early in the morning.

These hydathodes are the most common entry method for Xanthomonas campestris pv. campstris (which causes black rot). The leaf on the right is showing symptoms of black rot, with the lesion starting at the location of insect damage.

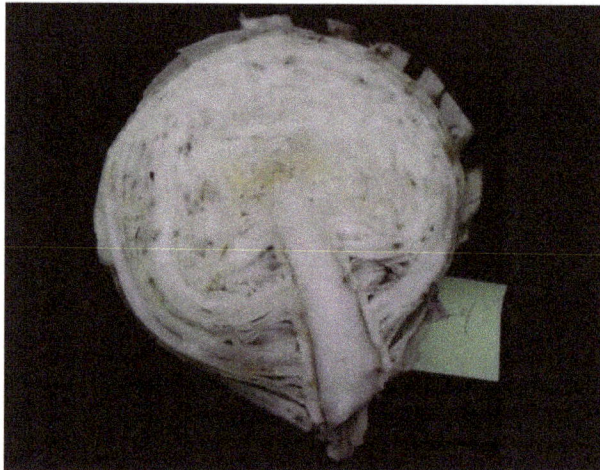

Internal vein blackening caused by the black rot pathogen. This head would rot completely during storage.

Prevention

Prevention is the best line of defense and is especially important in organic production. There are three preventative measures that can reduce the risk of a black rot outbreak.

Start with clean seed – It is known that the bacterium that causes black rot can survive on and in seed. Hot water treatment can be used to destroy the bacteria that may be infesting your seed. If you have purchased seed that has NOT been hot water treated, you can treat the seed yourself, but it is critical to do it correctly. For cabbage and Brussels sprouts, soak seed for 25 minutes in 122°F water; for Chinese cabbage, broccoli, cauliflower, collard, kale, kohlrabi, rutabaga or turnip, soak for 20 minutes in 122°F water. Mustards, watercress and radish are more susceptible to heat damage, and should be soaked for 15 minutes in 122°F water. Treat a small number of seeds the first time to ensure that the treatment is not reducing seed germination.

Use clean transplants – If you are growing your own transplants, make sure that the greenhouse has been cleaned well prior to starting transplants—even if you had no disease last year! Bacteria have a remarkable way of surviving in weeds, organic matter, or nooks and crannies, so if possible, get rid of all weeds, use new or disinfected flats, and disinfect benches and tools prior to the start of a new season. Be sure to keep foliage as dry as possible, and do not brush or trim wet plants. Use pathogen-free organic starting mix, and if you are adding compost, be certain that no diseased plant matter was used.

Rotate with non-crucifers – Because the black rot bacterium can survive in debris in the soil, it is important to rotate away from crucifer crops for a minimum of three years.

Reducing Disease Risk during the Growing Season

Anything that can be done to reduce leaf wetness and water splash will help reduce disease spread. This includes watering plants in the morning so that leaves dry prior to sunset, maintaining your irrigation system to reduce the likelihood of ponding, increasing spacing between plants, and orienting rows with prevailing winds to maximize air flow and drying.

Cabbage and cauliflower plants at this production facility are watered early
in the morning so leaves will dry quickly.

Management Strategies

As with most bacterial pathogens, managment can be very difficult when the weather is conducive to disease. Once a plant is infected, there is no rescue treatment since the infection is systemic. Copper-based products are effective in reducing spread from infected to healthy plants.

Before applying any pest control product, be sure to read and understand the safety precautions and application restrictions, and make sure that the brand name product is listed in your Organic System Plan and approved by your certifier.

Although black rot can be severe, following the prevention strategies described above will reduce the risk of this disease. Although the pathogen can survive on farms, we know that this is not the most common source of inoculum on farms that use a minimum three year rotation; instead, the pathogen is most commonly brought onto farms on seed or plants. In New York, new strains of the pathogen enter the state each year. Thus, planting only clean seed and disease-free transplants are the most important management practices in regions with cold winters. In locations with mild winter temperatures, the risk of maintaining the pathogen on farms is greater.

References

- Zadoks, Jan C. (16 October 2013). Crop Protection in Medieval Agriculture: Studies in pre-modern organic agriculture. Sidestone Press. ISBN 9789088901874. Retrieved 11 November 2017 – via Google Books

- Dovorak, P. "BLACK POLYPROPYLENE MULCH TEXTILE IN ORGANIC AGRICULTURE" (PDF). Czech University of Life Science Prague, Kamýcká. 52. Retrieved 16 November2014

- Tamburini, G., De Simone, S., Sigura, M., Boscutti, F., Marini, L. And Kleijn, D. (2016), Conservation tillage mitigates the negative effect of landscape simplification on biological control. Journal of Applied Ecology, 53: 233–241. Doi:10.1111/1365-2664.12544

- "Soil Compaction and Conservation Tillage". Conservation Tillage Series. Pennstate – College of Agricultural Sciences – Cooperative Extension. Retrieved 26 March 2011

- Crop-weed-management-strategies: gardenorganic.org.uk, Retrieved 21 January, 2019

- Wright, M. G.; Hoffmann, M. P.; Kuhar, T. P.; Gardner, J.; Pitcher, S. A. (2005). "Evaluating risks of biological control introductions: A probabilistic risk-assessment approach". Biological Control. 35 (3): 338–347. Doi:10.1016/j.biocontrol.2005.02.002

- Organic-crop-management-insect-management, organic-crops, crops-and-irrigation, agribusiness-farmers-and-ranchers, agriculture-natural-resources-and-industry, business: saskatchewan.ca, Retrieved 22 February, 2019

- Seaman, Abby. "Integrated Pest Management". University of Connecticut. Archived from the original on 20 February 2012. Retrieved 13 March 2012

- Disease-management-in-organic-lettuce-production: extension.org, Retrieved 23 March, 2019

- Early-blight-management-for-organic-tomato-production: extension.org, Retrieved 26 June, 2019

Permissions

All chapters in this book are published with permission under the Creative Commons Attribution Share Alike License or equivalent. Every chapter published in this book has been scrutinized by our experts. Their significance has been extensively debated. The topics covered herein carry significant information for a comprehensive understanding. They may even be implemented as practical applications or may be referred to as a beginning point for further studies.

We would like to thank the editorial team for lending their expertise to make the book truly unique. They have played a crucial role in the development of this book. Without their invaluable contributions this book wouldn't have been possible. They have made vital efforts to compile up to date information on the varied aspects of this subject to make this book a valuable addition to the collection of many professionals and students.

This book was conceptualized with the vision of imparting up-to-date and integrated information in this field. To ensure the same, a matchless editorial board was set up. Every individual on the board went through rigorous rounds of assessment to prove their worth. After which they invested a large part of their time researching and compiling the most relevant data for our readers.

The editorial board has been involved in producing this book since its inception. They have spent rigorous hours researching and exploring the diverse topics which have resulted in the successful publishing of this book. They have passed on their knowledge of decades through this book. To expedite this challenging task, the publisher supported the team at every step. A small team of assistant editors was also appointed to further simplify the editing procedure and attain best results for the readers.

Apart from the editorial board, the designing team has also invested a significant amount of their time in understanding the subject and creating the most relevant covers. They scrutinized every image to scout for the most suitable representation of the subject and create an appropriate cover for the book.

The publishing team has been an ardent support to the editorial, designing and production team. Their endless efforts to recruit the best for this project, has resulted in the accomplishment of this book. They are a veteran in the field of academics and their pool of knowledge is as vast as their experience in printing. Their expertise and guidance has proved useful at every step. Their uncompromising quality standards have made this book an exceptional effort. Their encouragement from time to time has been an inspiration for everyone.

The publisher and the editorial board hope that this book will prove to be a valuable piece of knowledge for students, practitioners and scholars across the globe.

Index

www.ingramcontent.com/pod-product-compliance
Lightning Source LLC
Chambersburg PA
CBHW061935190326
41458CB00009B/2743